# THE ILLUSTRATED ENCYCLOPEDIA OF
# Minerals & Rocks

By Dr J. Kouřimský
Photographs by F. Tvrz

Edited by
Randolph Lucas

Photographs 1 and 4 were taken with the kind permission of Náprstkovo Museum in Prague, photograph 3 with permission of the Museum of Applied Art in Prague, and photograph 5 with permission of the Ethnographic Museum in Prague.
Photographs 6, 7, 8, 9 and 10 show works from metals and gems by the professors and pupils of the Applied Art College in Turnov.
Photograph 51 is by K. Neubert, photograph 346 by Ing. Z. Borovec.
The majority of minerals pictured in this book come from the collections of the National Museum in Prague.
Page 6: Rock crystal — Dauphiné, France; actual size $12 \times 13$ cm.
Page 8: Mining in the Middle Ages; woodcut from the work of G. Agricola 'De re metallica', 16th century.

Text by Dr Jiří Kouřimský
Photographs by František Tvrz

Translated by Vera Gissing
Graphic design by František Prokeš

English version first published 1977 by
Octopus Books Limited
59 Grosvenor Street, London W1

ISBN 0 7064 0518 8

Printed in Czechoslovakia

3/11/03/51-02

# CONTENTS

# FOREWORD

Man has been aware of rocks right from the most ancient of times. He found many uses for them as raw materials, and was, at the same time, always conscious of their varied beauty. Both these factors influenced his own development. Man's knowledge of minerals had a deep effect on the standard of civilization: the beauty of the stones was an added inducement to work and explore this field; it enriched man's taste for beauty and influenced the development of human culture.

Apart from concentrating on these aspects, the main mission of this book is to provide, in popular form, a compact survey of basic data, as far as this is possible, in the field of mineralogy, and subsequently geology, for all those interested in nature's minerals. It is therefore primarily intended for mineral collectors, whose numbers in recent years have risen considerably.

The contents of this book are directed towards this goal. In the general section, the problem of the relationship between man and minerals is examined in some detail, and the basic mineralogical and geological terminology is explained. Though the latter may seem technical and dry, it is essential for any true enthusiast to know these terms. In the chapter 'Man and Stone', greater emphasis is laid on the beauty of rocks than on their technical uses. This is intentional, for the aesthetic effect of rocks does a great deal to stimulate the collector's interest.

The reader will find the identification section of the book very rewarding. He will find many interesting historical facts relating to individual minerals, and also details of their past and present significance for man. With some of the minerals, special attention is given to the history of their discovery and often an explanation is given of their name. This does not apply to all the minerals, only where such particulars ease identification.

The main physical and chemical characteristics of minerals are always given in marginal texts. A more thorough reference to them is made in the actual text only when a further explanation is essential. The same also applies to the survey of the most important deposits. In some cases only a description is given, in others additional characteristics are necessary.

The main concern of this book is to satisfy the interests and needs of mineral collectors and all lovers of nature. I should be very pleased if this book helped them on their path of discovery of the wonders of the mineral world, which can bring so much joy and satisfaction to a receptive observer.

Dr. J. KOUŘIMSKÝ

# MAN AND STONE

Man's natural thirst for understanding nature, for unveiling its secrets and exploiting its riches to his own advantage is one of the main factors which have determined the standard of culture and civilization from the dawn of the history until the present day.

The relationship between man and nature's mineral resources, the standard of his knowledge and his ability to make the fullest use of them are a significant measure of the extent of cultural development at any time. The link with rock must have been as significant for primeval man, who could turn rock into a tool or a weapon, for builders of pyramids, for men of ancient times, who could feast their eyes upon the Venus de Milo, as for 20th-century man with his ambitious concrete constructions.

This, however, is not the only bond between man and rock. From earliest times man has been aware of the aesthetic value of rocks and minerals. Rare stones were valued and turned into precious jewels, which throughout history have so often been the symbol of riches and power.

The essence of the relationship between man and nature's minerals remains therefore basically unchanged. Only the standard has altered, for it depends solely on man's ability to express his relationship with rock.

Primitive men, whose standard of living still resembled that of a wild animal, had no tools, even no clothing, and their needs were negligible. The first mineral of whose importance they were aware, was rock salt. They found the salt deposits when following the instinctive behaviour of animals. Later, rock became Stone Age men's raw material, and as they dug and worked the stone, men also learned to use their hands and brains.

During the Stone Age, men used rock for tools and weapons. At first they used chippings of rock such as limestone and sandstone. However, they soon discovered that it was more advantageous to use sharp-edged quartz, flints, hornstone, chalcedony, obsidians, or slate, which they could sharpen or chip. By the end of the Stone Age, men were already searching for minerals of rarer origin, heavier in weight and richer in colour, whose properties were fundamentally different from other minerals. They became aware of nuggets and leaves of pure gold in the alluvium of streams, and of chunks of copper ore with their green coating. A new era in the evolution of man was gradually approaching; men were learning to melt copper in kilns and to use it. When their attention was caught by tin ore in alluvium (cassiterite) and they learned about the properties of copper and tin alloys, they discovered that bronze, the alloy gained from these metals, had more useful properties than copper. This marked the birth of a new era in human history — the Bronze Age. The use of bronze spread from the valleys of the Nile and Euphrates to the Mediterranean region, and the cultures of the Phoenicians, Cretans and Greeks began to flourish. As copper and tin were expensive rare minerals, rock continued to be used for the manufacture of common tools. At the same time the importance of stone as a building material was growing. The first stone palaces were rising in the valeys of the Nile and Mesopotamia, and in Egypt limestone blocks of immense weight were used for building pyramids.

The Iron Age began only 3,000 years ago, because obtaining iron from ore presents difficulties. A far higher temperature is needed to melt it than for the manufacture of copper or tin. As long ago as the 4th millennium BC, men had discovered meteoric iron (Mexico, North America, Greenland), but did not yet know how to use it. Attention was probably drawn to iron ore by its conspicuous blood-red and yellow surface. The ore was melted with charcoal, to obtain a fairly substantial amount of iron as a cheap substitute for the rarer bronze. The discovery of iron metallurgy initiated an unprecendented advance of human activity in agriculture and crafts. Mankind was on the brink of a long period when education and economy were to prosper and blossom.

In the Middle Ages iron became the most widely used metal. Iron ores were at first extracted from pits, then from shallow mines. The tools of the Middle Ages miner were extremely simple, but gradually mining techniques improved. The old, exhausted deposits in the Roman Empire were progressively abandoned, and in the 8th and 9th centuries new mines were sought. An

1 Large emerald mounted in a piece of jewelry, used as turban decoration (Koseir, Egypt); Turkish work from the 17th—18th century.

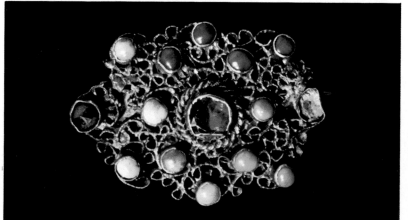

2 Filigree of gilded silver with almandine (Asia Minor) and with turquoise (Iran); Ottoman work from the 18th century.

3 Detail of gold malachite set (Urals); Viennese work from the beginning of the 19th century.

4 Gold bracelet with turquoise (Iran); Turkish work from the first half of the 19th century.

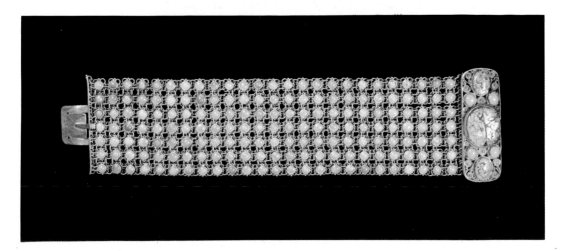

expansion in mining took place particularly at the time of Charlemagne in France, and in central Europe in Bohemia, Saxony and Slovakia. Silver too was mined there in large quantities; in many places silver was mined before iron. However, at the start of the Modern Age extraction of this silver was overshadowed by the import of cheap silver from newly discovered and extremely rich American mines.

As industrial manufacture flourished, the importance of minerals increased in all branches of manufacturing and industry. Minerals were important as ores, fuel, or raw materials for the chemical, glass, ceramic and building industries, and for the manufacture of china and heat-resistant materials. The use of mineral substances spread even into agriculture where they were used as fertilizers. With the advancement of time minerals were used to increasingly greater advantage, and mining methods have been continually improved right up to the present day. It is impossible now to imagine life without mineral resources, for they are inseparably connected with the economic and cultural development of man.

Apart from the practical use of rock and minerals, men have been fascinated, even in antiquity, by the beauty of some minerals. Men gazed at the shape of crystals whose origin they could not explain, and it is not surprising that sometimes they thought the crystals held supernatural powers. Even in the Stone Age, primitive men became interested in gold, which was easy to melt and to cast. From this the first pieces of real jewelry were made. Silver was mined along with copper during the Bronze Age for the same purpose. Jewels, ornaments and cult statuettes were also made by ancient man from less noble materials such as copper and bronze, or from rocks of unusually striking colour or shape. Exquisite ancient gold jewelry has been found in regions where the metal was in abundance, for example in Peru and Mexico.

Precious stones were first used in prehistoric times but it is impossible to decide which nation worked them first. The oldest reports come from India and from countries of the eastern Mediterranean, where their deposits were found. Other well-known deposits are in Sri Lanka (rubies and sapphires), and on the island of Zebirget in the Red Sea (olivines). The engraved gems on the Babylonian-Assyrian seal rolls originate from the 4th millennium BC. The Babylonians excelled in cutting rock, as can be seen on the large cylinders of carnelian or hornstone drilled through their centres and used as amulets.

The use of precious stones spread from the Orient to Greece, where the art of gem engraving — glyptography — originated. The names of artists of this era have survived through the ages. The perfection of the work on precious stones in Greece led to the first secular objects made from them. The technology was similar to that of today. Grinding was done on grindstones, or on hand-operated wheels; powdered emery or diamond was used for an abrasive.

The Romans' knowledge about precious stones grew and grew. They acquired working methods from the Greeks and gathered further knowledge from the nations of Asia and North Africa. The oldest gems were precious stones which were cut inwards, *intaglio*, or in a raised form, *cameo*. They have been found in Etruria and show the influence of Egypt. The Roman stone-cutters were mostly Greek in origin, and they founded many workshops, especially in southern Italy. The Romans' knowledge of rocks was far superior to that of the Greeks. The first work of Pliny (died AD 79) on natural history gives details of their deposits and describes crystal structures. Pliny also mentions that the Roman Emperor, Nero (AD 37—68), used an emerald spectacle lens. Often in those days the identity of stones was mistaken and there are reasons to believe the lens was actually made of aquamarine. All the same, Pliny's reference is valuable because it is the first reference to the technical use of a precious mineral. After the disintegration of the Roman Empire, glyptography moved, mainly to Byzantium.

In the early Middle Ages ownership of precious stones was confined to rulers, high nobility and dignitaries of the Church, particularly in Europe. The popularity of precious stones fell after the disintegration of the Roman Empire. This was due, to a large extent, to the new philosophical trend, which viewed the world and its values in a different light. Early Christianity, in comparison with previous philosophical movements, turned from life on this earth to life after death. To be adorned with jewels, which seemed so natural to the heathen, especially in the highly cultivated ancient world, was now considered an extravagance. During that time many deposits of precious

12

stones stood neglected. It appears that in the early Middle Ages many methods of working stones that had been commonly used in earlier times were forgotten in central and western Europe. For instance, stone engraving, making gems and cameos, which were popular processes in the ancient world, with echoes dying out in Byzantium, became almost unheard of. We know this because in the seals of great men, cut minerals were used from previous ages, mostly from former Roman regions, but also from India and other distant lands. Some ancient gemstones were even used for making Christian objects of worship, and on many occasions Greek and Roman subjects were used to represent events from the Old Testament.

In the Gothic Age attention was centred primarily upon rough work on transparent and translucent gemstones. These were never used in their natural state. The stone-cutters removed imperfections from the mineral and gave the stones at least a superficial polished finish. This process remained rather primitive for a long time, not only because of the poorness of the technique, but also because an effort was always made to preserve as much of the mineral as possible.

The expansion of the arts all over Europe during the 13th and 14th centuries triggered off a new interest in glyptography. The first signs of this were in France, then gradually in other European countries. By 1290 an engravers' guild had been formed in Paris whose craftsmen made an effort not only to follow the methods of their predecessors, but also to create independently. This was strongly supported by French and Burgundian rulers of the 14th century, who founded whole collections of cut and polished antique stones, especially gemstones. There was a renewed interest in trading with gemstones, and large cities, through which routes from the eastern stone deposits led, became centres of importance. Baghdad and Cairo were the most noted ones. These places furnish us with written information even about the value of precious stones in the middle of the 13th century. It is impossible to convert them to current prices, because the way of life and the standard of living during that time are not comparable with the present day. All the same, it might be of interest to mention as an example, that a beautiful ruby lens, weighing 3 grammes (15 carats) was valued in Baghdad as 300 grammes of gold. It shows that the value of precious stones was extremely high even at that time.

Precious stones, whether polished or in their natural state, were at that time sold mainly according to weight. For the more valuable ones a special weight unit — a carat — had been used through the ages.

It is said that the name 'carat' originates from the weight of a carob seed (*Ceratonia siliqua*), whose pods were imported to Europe as 'St John's bread'. The carob seed, from trees of the Mediterranean regions, has actually been used for weighing precious stones from time immemorial. The weight of each seed of 'St John's bread' is usually the same, being about one-fifth of a gramme. Arabs called these seeds 'kharrub', from which the Greek word 'keration' was derived, then eventually the English 'carat'.

Originally the carat was divided into halves, quarters, eighths, sixteenths, thirty-seconds, even sixty-fourths. In addition, the ounce, which had 144 carats, came into use in France as a special weight unit. The carat was commonly used, but its exact size varied from place to place between 0.197 and 0.216 grammes, which led to great confusion. If someone, for instance, bought stones according to the carat valid in Florence, and sold them in Livorno, he gained as much as 95 carats on each 1,000 carats!

This matter was not put in order until 1907 when the so-called metric carat, of exactly 0.200 grammes, i.e. one-fifth of a gramme, was introduced in France, Italy and Germany. It did not become a truly international unit until the mid 1920s.

The most noted treatise on the subject of precious stones from the 13th century comes from the pen of an Arabian merchant, Ahmed ben Jusuf al Teifassi, and gives the values of precious stones in markets of the Near and Middle East. This work is of great importance for various reasons. First, it gives a complete list of the precious stones sold at the local markets. Here we can find both their relative and absolute value, and can compare ben Jusuf's survey of 1242 with current world price lists of gemstones.

The minerals offered for sale, and listed according to the relative values of that period, are as follows: rubies, emeralds, diamonds, spinels, cat's eyes, blue sapphires, zircons, yellow sapphires,

5 Amulet with precious stones; national work of the Balkans from the end of the 19th century.

6 Figure from tiger's-eye (Republic of South Africa); 5 × 3 cm.

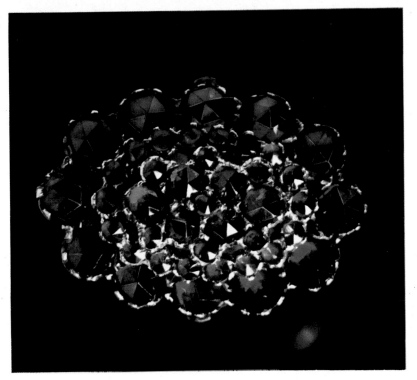

7 Gold brooch with Bohemian garnets (Central Bohemian Highlands); 4 × 3 cm.

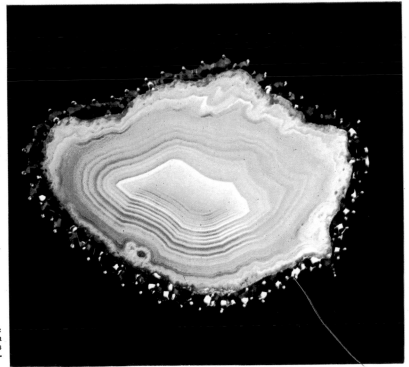

8 Agate brooch (Krkonoše Mountains, Bohemia) with a frame of Bohemian garnets (Central Bohemian Highlands); 5 × 3.5 cm.

beryls, turquoises and a type of garnet (almandine). The most valuable gem, the ruby, fetched as much as 2 to 6 gold dinars per carat. Its value was based on the quality of the stone and on the size. The emerald, on the other hand, had a uniform price, 4 dinars per carat. This uniform valuation, which is in direct opposition to today's prices of emeralds, was apparently caused by the rich emerald mines of upper Egypt yielding a sufficient flow of large minerals. The price of diamonds fluctuated between 2/3 and 4 dinars. Spinels had only half the value of the lowest quality rubies. Blue sapphires and zircons were valued at a quarter of the price of spinel.

It is, unfortunately, impossible to tell from Ahmed ben Jusuf's work in what form and quality diamonds were sold. We assume they were of high quality, but in their natural form. All other stones were cut to an irregular lenticular shape, as was then the general custom. Many had holes drilled through them so that they could be used for jewelry. This was the normal method used throughout the Middle Ages for working transparent and semi-transparent precious stones.

It has already been stated that in the 14th century precious stones were primarily used for religious purposes and to emphasize the power of rulers and of the high feudal nobility. But the market for ornamental objects and pieces of jewelry made from precious stones was slowly spreading to other social classes. Quartz (rock crystal) was becoming the main mineral in use. It took the place of glass for protecting valuable objects, or was used for the manufacture of lenses. Sometimes metal foil was placed under the ground quartz, so the jewels would give an impression of magnificence and wealth. Quartz vessels encased in silver were often the property of rich bourgeois families. In those days many households kept whole lumps of natural quartz.

It is obvious, from all that has already been said, that in the Middle Ages precious stones were widely used. But the theoretical knowledge of them was rather superficial in comparison with the knowledge of the ancient Romans (Pliny for example). Several 'theoretical' works on precious stones by medieval authors mention many superstitions about the supernatural powers of some gems. The dogmatism of the Middle Ages did not naturally favour the study of the laws of nature. It follows therefore that even books written at the end of the Middle Ages mostly repeat the findings of previous authors.

The most frequently quoted book from the Middle Ages is the work of Bishop Isidore of Seville (c. 570 – 636). An ecclesiastical writer of many informative and historical works, he provided for the Middle Ages an encyclopedia of knowledge with his *Etymologiae* (Beginnings), written strictly within the Catholic ideology of the time. Another important medieval work was the book of Constantine Psellos (1018 – 78), a Byzantine statesman and historian, who also studied natural science. His book dealt primarily with the healing and magical powers of gemstones. This book like other works of the time on precious stones, drew upon the information provided by ancient writers, especially Pliny, modified, and often clouded, by the prevailing ecclesiastical dogma.

At that time gems called 'Moon-stones' were very much in favour, for they were supposed to carry strange powers, influence health, physical beauty, bring wealth, glory and happiness. These stones were worn as amulets and many were used in the preparation of medicine.

A manuscript which deserves closer attention is by an author, who, unlike the writers mentioned up to now, tried to pass on to his readers some of his own conclusions and observations. He is not purely concerned with repeating old facts and quoting the official view of the dogmatic philosophical opinion. The author of this work is Georgius Agricola (1494 – 1555), a renowned surgeon and mineralogist, who lived in Saxony and Bohemia. He explores the subject of precious stones in one of the volumes of his widely acclaimed book *De natura fossilium libri X* (Ten Books on Stones' Character).

In the 15th and 16th centuries glyptic art spread further still. This happened when Constantinople was captured by the Turks and many local artists emigrated to other lands, especially Italy, where cities such as Florence, Rome, Naples and Venice gradually became important centres of glyptography. As in many other branches of artistic and creative endeavour at that time, Italian work in glyptography influenced the development of the craft in other European countries.

At that time, people working precious stones concentrated mainly on making ornamental vessels, jewelry and intaglios. Amber, found on the shores of the Baltic, was a widely used mineral, and many exquisite pots and jugs made from it, often with relief figures, have been preserved to this day.

Varieties of quartz, especially from Saxony, Silesia and the Bohemian Giant Mountains, were also very popular.

Of course, interest in the smaller, but more valuable stones, which were much in demand during the Middle Ages, did not die out. Diamonds began to be used on a large scale, though often other clear minerals, mostly the perfectly transparent crystalline quartz, hid behind the name of 'diamond'. Old manuscripts inform us that sapphires, Persian turquoises, Egyptian emeralds, rubies and 'their poorer associate — spinel' were also very much in use.

During the Renaissance period, the sequence of values of precious stones remained basically unchanged. Their relative valuations altered, however, for the price of the more valuable stones rose more rapidly than the price of cheaper stones. Information about precious stones used at this time is given by Benvenuto Cellini (1500—1570) in his memoirs. He was a noted Italian goldsmith, medallist and sculptor, who was active in Florence, Rome and France.

In the 17th century, Prague became an important centre of glyptic art. It is therefore not surprising that essential information on precious stones can be found in a book *Gemmarum et lapidum historia* (The History of Precious and Common Stones), published in Prague in 1609. Its author, a Dutchman by origin, was A. Boëtius de Boot, a personal surgeon of the Emperor Rudolf II and a versatile scientist.

In the 17th century the diamond became pre-eminent among gemstones. This happened because the glyptographic technique was gradually being perfected. New types of cuts were used, especially the brilliant cut. This, apparently, was suggested by a famous French politician and a lover of precious stones, Cardinal Mazarin (1602—1661). By cutting a diamond into this shape, the lustre and the play of colours were displayed to their greatest advantage, whereas the rosette cut, used from the start of the 16th century, was duller in comparison.

9 Crystal intaglio (Madagascar); 39 × 26 cm.

10 Carnelian cameo (India); 19 × 15 mm.

11 Rock crystal pendant and a silver ring with crystal; contemporary work.

12 Natural sapphires in clockwork; Viennese work from 1880—5.

13 Synthetic ruby, raw material for the manufacture of a laser; diameter of pin 24 mm.

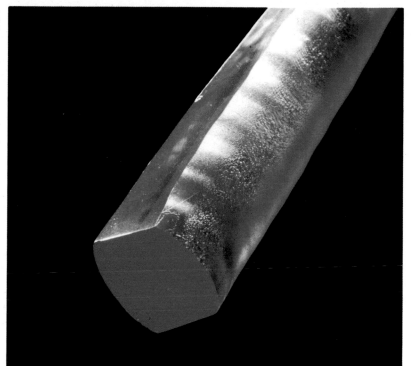

14 Pure gold bricks; total weight 72 kg.

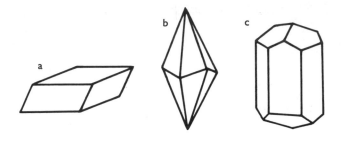

Various characteristic forms of calcite crystals:
a) rhombohedron, b) scalenohedron, c) columnar
crystal

Since the 17th century, rosette and brilliant cuts have been the most commonly used cuts on all transparent precious stones. When we discuss a brilliant, it is necessary to stress that this is not a kind of diamond, as many seem to think, but a cut into which many other minerals can be shaped. When cutting the gem, the top and lower facets are cut in such a way that the lustre and colours shine to their full capacity. On the upper surface of a brilliant the facets lie in two, three or four layers over each other. A brilliant is known accordingly as two-layered, three-layered, etc. Diamonds which are very small or imperfect, and those less transparent are even today cut into the rosette shape. This cut has facets on the upper surface only. The base is completely flat.

When these cutting and grinding processes are used, less care is given to trying to preserve as much of the stone as possible. Instead, attention is concentrated on making the stone's appearance as perfect as could be, which of course depends on the quality of the stone, its lustre and colour.

The 17th century was not only important for the developments in glyptography and for concentrating on the creation of objects of high artistic value. It was a period when alchemy flourished, when people tried to transmute base metals into gold and searched for an 'elixir of eternal youth'. It was also a period of discovery of the laws of natural science, a period when the science of chemistry was born.

Before chemistry was established as a science, alchemy and 'iatrochemistry' (medical chemistry) were significant movements. They tried to find a chemical explanation for the evolution of life and also to treat illnesses chemically. Many minerals and alloys were used for the preparation of medicines. So it is not surprising, even at this time of the birth of true scientific chemistry and medicine, that particular attention was focused on gems, which were believed to have supernatural powers.

During the second half of the 17th century, monumental Baroque buildings were constructed in most parts of Europe. They emphasized the power of rulers, the Church and the feudal nobility, who all lived in stately luxury. People were not really interested in real values. New fashions were created and they demanded such vast quantities of so many types of stones, that the supply of real gems was completely unable to satisfy the demand. Glass imitation stones were therefore often used for making jewelry. The Renaissance may have been a period of expansion in the use of precious stones, but now there was a definite decline. Diamonds, rubies, sapphires, emeralds, amethysts and clear quartz were practically the only stones to be used.

Not until the second half of the 18th century did society start to look at precious stones in a different light. After the French Revolution, a new class of society rose to the fore which became interested in jewels and in valuable objects not only for its own personal adornment, but also for improving its life standards. The glyptographic craft kept in step with the rise in prosperity of the new social order, and blossomed anew. The ever-growing demand for precious stones necessitated the search for fresh deposits and for discovering new types of gemstones.

In the last decades of the 19th century and during the 20th century, the popularity of precious stones was, to a large degree, determined by fashion. Attractive, natural gems were very much liked, and high prices were paid for them. The success of the synthetic production of some gemstones caused a temporary threat to the real stones. Particularly in the 1920s, many people wore

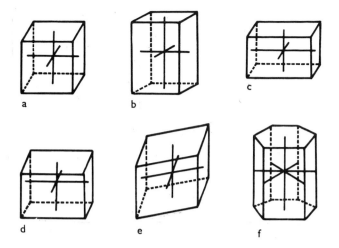

Axial crosses of the crystal systems: a) cubic system, b) tetragonal,
c) orthorhombic, d) monoclinic, e) triclinic, f) hexagonal, or rhombohedral

so-called imitation jewelry, which certainly harmed the sales of the more expensive natural gems. It did not take long, however, before the popularity of the natural stones returned, and their value began to climb once more again. This was only to be expected. After all, the genuineness of the stone does not appear only in its value, but also in its appearance. Even today, in spite of modern techniques and the successes of the synthetic chemist, however perfect an imitation stone may be, it does not have the physical properties of a genuine gem.

These few pages have explained how, over the centuries, the relationship between man and minerals developed and strengthened, and some examples of how man progressed in utilizing these resources for his own needs have been given. At times it seems that men have perhaps been over-eager to master mineral wealth. This was particularly true when they felt that wealth was almost within their grasp, that all they needed to do to make a fortune was to find new mineral deposits. In many corners of the earth great expeditions set out, hoping to discover new deposits of precious metal, precious stones and other raw materials. This 'gold fever' brought hard toil and battles with Nature in the quest for her treasures, but also, alas, merciless fighting among men which claimed many human lives.

During ancient Egyptian times campaigns set out in search of gold, rich metal ores and deposits of precious stones and other raw materials. The efforts of the primeval nations to find flint, 'the strategic material' of that time, and the expeditions of ancient Romans to the Baltic coast amber deposits are both well known.

The largest expedition of this kind, motivated greatly by the lust to find gold, silver and precious stones, resulted in the colonization of Central and South America. The colonizers slowly penetrated into the heart of the newly discovered continent around the year 1509, and in just a few years took over the Inca Empire (from Mexico to Peru) with its gold-mines, Colombia with its emerald deposits and finally the territory of present Brazil, rich in gold and topaz. The settlers wasted no time in extracting gold in Central America and emeralds in Brazil; they had already started work there by 1537. The importation of Brazilian topaz to Europe caused a worldwide fall of the price of this stone.

In 1883 gold fever erupted on the banks of the northern Chinese River Zeltuga (a tributary of the Amur), where gold deposits, said to be the richest at that time, were discovered. Inhabitants from the neighbouring regions of Russia and China sped to the scene, and so did many other adventurers. The gold-miners formed a large colony, which had as many as 20,000 inhabitants. They proclaimed themselves as the independent republic of Zeltuga and laid down their own laws.

15 Structure of rock salt crystal floating on salt lake (Fayum, Egypt); 4×3 cm.

16 Contact colour radiogram of a uraninite vein (Jáchymov, Czechoslovakia); 7×4 cm.

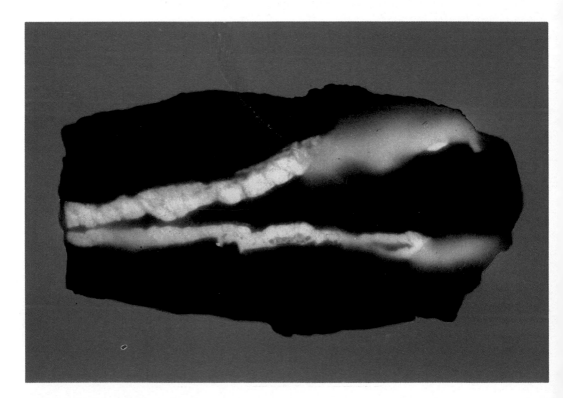

But the influx of foreign prospectors made the Chinese government feel uneasy and they liquidated the colony in 1886 by force.

The biggest 'gold rush' in the history of man followed the discovery of gold on the banks of the Yukon in the Klondike in northwest Canada in 1896. Over 100,000 prospectors set out with a single goal in mind — to become rich as quickly as possible. Only 4,000 of them were fortunate enough to find gold, and for merely a few the journey proved worthwhile.

'Diamond fever' did not lag far behind. The most famous diamond rush came in 1867, when the first diamonds were discovered in South Africa. The magazine, *Diamond Fields Advertiser*, from Kimberley said at that time: 'In ports sailors abandoned ships, soldiers fled from the army, policemen discharged prisoners, merchants rushed from their shops and clerks from their offices. Farmers turned their back on their herds and all of them rushed to the banks of the Vaal and Orange rivers to dig for diamonds.'

There is quite a colourful and dramatic story attached to the discovery of large diamond-bearing deposits in Yakutsk. In the 1930s, Sobolev, a geologist from Leningrad, arrived in Yakutsk. After extensive groundwork during several weeks he began to compile a geological map of the Siberian Plateau. During the course of his work he came to the conclusion that there was a certain analogy between this region and South Africa. He came forward with the hypothesis that Siberia must also be rich in diamonds. Not till after the end of the Second World War, in 1947, did an expedition of geologists leave Irkutsk and after many adventures in the wild, uninhabited taiga, they found the first diamond in 1949.

Later discoveries of mineral resources tell a similar story; today extensive research is first carried out by scientists in the field and in laboratories. An example of this was the discovery of large

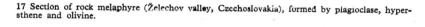

17 Section of rock melaphyre (Želechov valley, Czechoslovakia), formed by plagioclase, hypersthene and olivine.

Structure of hexagonally symmetrical water crystals (snow flakes)

Asymmetrical crystal of quartz

deposits of metal and uranium ores in the Congo, of petroleum in Libya and in other countries and of a variety of mineral deposits found in many developing countries. New methods for prospecting have been worked out, utilizing the most modern mechanical techniques.

The documentary material, acquired by collecting samples on the actual terrain, must form the basis of all experiments. This is the most important role of the mineral and rock collections amassed in various museums and research institutes. Old collections which once belonged to rulers and the nobility form the core of many such exhibits. The collections have been enlarged by aubsequent additions from research institutes or from the private collections of different researchers snd collectors. In this way amateur mineral collections can be of value for new scientific endeavours.

# DEFINITION OF MINERALS AND GLOSSARY OF TERMS

The stone with which we are most familiar, because it often forms whole cliffs and is also used as a building material, is a rock and not a mineral. If we examine its uneven surface, especially through a strong lens, we will see that it is composed of a multitude of large and small grains of various colours and shapes. If we take a close look at sand, we can see similar grains. Sand grains consist mainly of colourless quartz and mica flakes. Quartz and mica are minerals, but sandstone or a cliff boulder are rocks.

A mineral is a homogeneous inorganic natural substance of a definite chemical composition, usually a solid, but occasionally a liquid. A few are composed of a single element, but most are formed by the combination of several elements. A product which has developed by a natural process is a natural substance. A mineral is homogeneous physically and chemically when it consists of elements of the same kind throughout. In this a mineral differs from rock, which is also an inorganic natural substance, but one composed of different minerals, and which, in contrast to them, forms large parts of the earth's crust. There are some rocks (for instance sandstone or limestone), which consist almost entirely of one mineral (quartz or calcite for example). We do not classify them as minerals, for they do not have to be homogeneous, as do minerals.

The science which studies minerals is called mineralogy. It not only observes and defines the various characteristics of minerals, but also explains their origin and development. Petrography concentrates on studying rocks. These sciences, together with geology, which studies the origin, composition and evolution of the earth, are not just theoretical sciences. Their findings affect practical life with ever increasing force. The vast and vital role that minerals and rocks play today is taken for granted. They are the basic raw materials of industry, whether they are metallic minerals extracted from ores, minerals for the chemical, glass or ceramic industry, or others.

24

# MINERAL STRUCTURE

The majority of minerals and other solids can, under certain conditions, occur in regularly shaped pieces called *crystals*. The regularity of shape and the flawless smoothness of the lustrous faces of crystals fascinates most people at first glance. Crystals vary greatly in shape and size. Some are minute, visible only through a microscope; others may measure several metres. The shape of a crystal can be as thin as a needle (acicular), or columnar, tabular, fibrous or lamellar.

A closer look at crystallized minerals shows that their crystals are structurally nearly always the same. It is then possible to conclude that a mineral has a definite crystal structure, which is not accidental but is determined by certain laws. The science which pursues the study of these natural laws and explains and describes crystal structures of different minerals is called crystallography.

Natural scientists have studied crystal structure from days of old, but crystallography gained recognition as a true science only in the 17th and 18th centuries. Crystallography, as we know it today, is based upon the findings of many other scientific fields, especially mineralogy, chemistry, physics and mathematics.

The basic difference between crystalline and non-crystalline matter does not lie only in the regularity of arrangement of the external faces. After all, crystalline grains in rock minerals often have irregular shapes and yet they are crystals. The basic difference is in their internal structure, which means the arrangement of the smallest particles — molecules, atoms and ions. These tiny particles are chaotically arranged in gases, liquids and non-crystalline solids. In crystals, however, they have a regular, repeating pattern.

As has already been mentioned, some minerals crystallize in a definite characteristic structure. But the structure of many crystallized minerals varies. Calcite crystals, for example, occur in various forms: high or low rhombohedra, columnar, acicular or tabular forms. These shapes share one characteristic, however; they are symmetrical, and their symmetry is in line with the group symmetry of the smallest particles of the substance.

According to this symmetry, crystals are divided into seven major groups, called crystal systems: 1) triclinic, 2) monoclinic, 3) orthorhombic, 4) tetragonal, 5) trigonal, 6) hexagonal, 7) cubic. Crystals are classified as belonging to a particular system according to their axes of symmetry which are, basically, imaginary lines passing through the centre of a crystal, and also according to their number and kind. All crystals belonging to an individual system must have a certain characteristic form, which corresponds to the relative symmetry. This means, in fact, that each crystal system has its own individual crystal form.

The crystal structure of every mineral is determined mainly by its chemical composition and by physical conditions during its development, especially temperature and pressure. Crystals with completely even faces do not often occur, for the various external conditions make such perfect growth of faces impossible. The size and the development of the individual crystal faces are not so vital for classification as the angle between faces, which is the same for every crystal of the same system. Crystallography is largely based on the study of these interfacial angles. Instruments used for measuring the angles are called goniometers.

*Distortions of crystal faces*, which develop either during the growth of a crystal, or through the effects of weathering, can be an aid in identifying a mineral (the faces of pyrite crystals are, for instance, often striated). The conditions of environment affect the speed of growth of crystals, and also their physical properties. When a cluster of crystals grows in a confined space, they crowd each other and impede each other's development. If a growing crystal becomes entangled with a neighbouring crystal, it cannot continue to grow in the original direction, but can expand in other directions. This is why the grains of rock have an irregular shape. On the other hand, crystals which originate in a soft, yielding environment, such as volcanic tuff, or in certain sediments, are often able to develop perfect faces.

Crystals also form in rock crevices, where they are deposited either by solutions or gases. Here the conditions for development are very different and much less favourable. The growing crystals cannot develop faces upon the rock surface to which they adhere. They can continue in their growth only towards the centre of the cavity. Clusters of crystals which grew from a common base in fairly parallel lines are called *druses*. Such parallel growth can occur only if the wall of the crevice is fairly even. If the rock crevice is round, the adhering crystals protrude with their free ends towards its centre, and are called *geodes*.

It is common for minerals to occur grouped in druses and geodes. They are frequently found in ore-veins and in cracks and crevices of the most varied types of rocks. Quartz, calcite, fluorite, barites and similar minerals are often grouped in druses. In some large cavities and crevices truly magnificent crystals can develop, as can be seen, for instance, in the 'crystal cellars' in the Alps.

The assemblage of crystals in druses and geodes is usually in a fairly regular pattern. But if the crystals grow in an interlocking, irregular clump, it is called an *aggregate*. These aggregates are often made up of multitudes of tiny crystals, more often than not with faces imperfectly bounded.

Apart from aggregates there are also mineral clusters which are basically made of parts of individual crystals not visibly bounded externally. The minerals which usually appear in this form are those which commonly develop large crystals, such as quartz.

For some irregular or round crystal formations in rocks the term *nodule* is used in mineralogy and petrography. Usually of quartzitic composition, nodules are often found in limestone beds. Chalcedonic and opal varieties also develop nodules, which originate through deposition of marine plankton.

It has been stated that the growth of individual crystals is governed by definite laws. These laws can often be applied to whole groups of crystals. Sometimes two or more crystals of the same chemical composition grow together with definite rules, governing their relative positions. Depending on the number of the twinned crystals, we talk of *twins*, *triplets*, etc. The intergrowth of two different minerals, though perhaps of the same chemical composition but with a different internal structure is not often found in nature (pyrite and marcasite, or even some minerals with a completely different chemical composition, such as haematite and rutile, are some examples).

Another interesting mineral form is a so-called *pseudomorph* or *mineral mimic* which is frequently found in the mineral kingdom. For instance, limonite, which has no crystal form of its own, is sometimes found in the shape of a perfectly bounded pyrite crystal. This unusual happening occurs because the original crystal of pyrite has become decomposed through the effects of water and has been replaced by limonite. In other cases one mineral changes into another mineral of identical chemical composition. For instance aragonite ($CaCO_3$) has an orthorhombic crystal form, and it changes slowly and gradually into rhombohedral calcite, but the original crystal form remains unchanged. Sometimes a pseudomorph is the result of *incrustation* which is produced by a powder-like coating of one mineral being deposited on the crystals of another.

It is rarer to find in nature minerals which have no traces of crystalline structure and which are said to be *amorphous* (opal and amber are examples).

# PROPERTIES OF MINERALS
# AND THEIR IDENTIFICATION

Mineralogists, especially the ones who specialize in precious stones (gemmologists), are often approached by people with a request to identify their minerals, which may be in their original natural form, cut and polished, loose, or set in a piece of jewelry. People are often surprised if they are not immediately given a positive identification. It is by no means easy to identify a precious stone, especially if it has been cut and polished, and to verify its genuineness. Not even the most widely experienced specialist could risk identifying with absolute certainty a precious stone, especially one that has been worked on, just by looking at it. An accurate identification cannot be based only on appearance, i.e. colour, degree of lustre, play of colours, etc. In most cases a magnifying glass, or some other simple instrument does not suffice. To be absolutely safe, often a whole range of specialized apparatus is required.

It is vital to say a few words about the properties of minerals, as without such knowledge identification of minerals would be quite impossible. Generally speaking, the internal structure of a mineral determines its external form and other important physical properties. These may vary according to the character of the crystal form. Experiments have also proved that some minerals show different physical properties in different directions. It is essential to mention the most typical physical properties, which are an important aid in mineral identification.

Finding a cleavage of a mineral, for instance, can be most useful. *Cleavage* is basically the tendency to split along certain definite planes in relation to the crystalline form and internal structure and is possessed by many minerals. Cleavage often appears as small, hardly visible splits even inside the stones. Such tears, which reduce the value of precious stones, are often caused by mechanical damage. The perfection of cleavage depends on the mineral. Cleavage is therefore described according to the degree of its perfection; *eminent* (gypsum) with ideally smooth and straight cleavage planes, *perfect* (calcite), *good* (fluorite), and *poor* (garnet), when only uneven planes occur. An engraver must know the cleavage of all precious stones. Topaz, for instance, can be split only in one direction, a diamond in several directions, whereas quartz does not cleave in any particular direction.

Another important characteristic of a mineral is *hardness*, which is basically the extent of its resistance to mechanical injuries. For individual minerals hardness is determined by the well known 'Mohs' scale of hardness'. This has ten grades, and their order, from the softest to the hardest, is as follows: talc, rock salt, calcite, fluorite, apatite, orthoclase, quartz, topaz, corundum, diamond. The hardness may be tested by trying to scratch the mineral specimens of the scale with the mineral under examination, but never the other way round, to avoid damaging the examined mineral. Endeavouring to establish hardness, though it seems easy enough, creates some difficulties. It is impossible, for example, to establish an accurate hardness for earthy minerals, powdery, needle-

Crystal structure of a) graphite, b) diamond

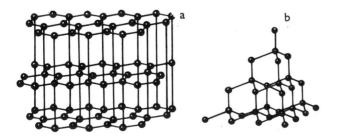

shaped, fibrous and lamellar aggregates, because blank spaces between individual grains, fibres or pores usually make the mineral seem to be softer. Sometimes it is the other way round — as with nephrite for instance, whose exceptional firmness, caused by interpenetration of microscopic fibres, seemingly increases the hardness. This method of determining hardness is not to be recommended where cut and polished precious stones are concerned, though laymen like to use it. A file often happens to be their tool, and 'identification' then may result in the stone being scratched and therefore losing value.

As has been mentioned, Mohs' scale of hardness is for grading purposes only, so the grades 1 to 10 do not actually tell us how many times any of the specimen minerals in the scale are harder than talc (which represents its first grade). Complicated, highly specialized apparatus called sclerometers are used to determine the absolute hardness. Equipped either with a diamond or a steel point, which is pressed into the mineral under examination, they measure hardness by the force needed to indent a specimen with the point.

The absolute hardness of a mineral is very different from the grade of hardness established by using Mohs' scale. Diamond, for example, is 4,000,000 times harder than talc, corundum 33,333 times, etc. The smallest differences in hardness are between grade 4 and 5 (fluorite — apatite), and 6 and 7 (orthoclase — quartz).

Another, perhaps even more important method of mineral determination, especially when testing cut and polished stones, is to find their *specific gravity* (also known as 'relative density'). This is a number indicating how many times the tested matter is heavier or lighter than an equal volume of water.

*Luminescence,* though never an entirely reliable method of determination, is certainly an interesting one. Certain minerals, when subjected to short-wave radiation, emit light. Ultraviolet lamps or X-ray equipment are used for this radiation test. Luminescence is particularly characteristic of certain minerals.

*Optical properties of minerals* represent vital physical characteristics. Some may be assessed with the naked eye, others have to be determined with the aid of instruments. The optical properties of minerals are determined mainly by seeing how a mineral transmits and refracts incident rays of light. Colour and lustre are the most obvious optical properties. But identification of a mineral and determination of the genuineness of a precious stone cannot be based solely on colour and lustre. Many minerals are, after all, identical in colour. It is therefore always advisable to carry out additional tests of other properties so that the identification of the mineral can be conclusive. Optical properties, when determined by appropriate apparatus, can give several accurate identification characteristics.

According to the degree of transparency we distinguish transparent, translucent, non-transparent, and opaque minerals (in the last case their thinnest sections are not transparent even when viewed under a microscope). In some minerals (such as pyrite) the transparency is constant whereas it varies in most of the others.

*Lustre* of minerals is divided according to intensity and quality. The strongest lustre is found in

Various crystal twins: a) gypsum, b) orthoclase (so-called Carlsbad twin), c) aragonite

  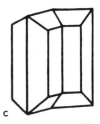

a                    b                    c

transparent and translucent minerals, and is called diamond (adamantine) lustre, typical of diamonds and certain zircons. The high lustre of opaque minerals is called metallic lustre. Other minerals have lustres which can be vitreous, resinous, greasy, dull, pearly, silky (common in fibrous aggregates), etc.

Lustre intensity depends entirely on the degree to which a mineral bends light rays which fall upon it and pass through it. The measure of the extent to which a substance bends light is called the *refractive index*. Its exact determination requires specialized apparatus such as a polarized microscope and refractometers.

The simplest way to find the approximate refractive index of transparent and translucent crystals, or cut precious stones, is as follows; the mineral should be placed in a transparent glass vessel filled with an immersion liquid of a known refractive index, and an intensive light should be focused upon it from above. Watch the shadow thrown onto a sheet of paper placed underneath. In general, except for certain crystals whose form resembles parallel plates bounded by perpendicular faces, crystals, or cut precious stones, with a higher refractive index than the liquid will throw a dark outline onto the paper and all convex edges outside will have a light outline. Objects with a lower refractive index throw a light outline, and the edges outside the circumference of the crystals are dark.

Examples of the varied cleavage of minerals: a) cubic cleavage of galena, b) rhombohedral cleavage of calcite, c) cleavage of mica running in one direction only

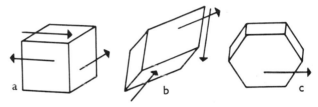

18 Biotitic granite — Mrákotín (Czechoslovakia); 10 × 6 cm.

19 Pegmatite — Etta mine (South Dakota, USA); 11 × 8 cm.

20 Gabbro — Minnesota (USA); 12 × 9 cm.

21 Anorthosite — Labrador (Canada); 12 × 10 cm.

The effect becomes more pronounced if the vessel is gradually moved away from the plane of projection. Its intensity depends also on the optical density of the tested object and of the immersion liquid. If the mineral and the immersion liquid have the same refractive index, the edges begin to fade away. This method can be used to determine the identity of precious stones which are rather similar, or to distinguish glass from genuine gems.

*Colour* of minerals may be determined by intrinsic colour, due to chemical composition (malachite for instance) but in other cases it can be often affected by the inclusion of foreign chemical substances (with spinel or topaz for instance). It can be influenced mechanically by the intrusion of a foreign pigment, as happens with haematite in ferruginous quartz or in carnelian. From the optical point of view the colour of a mineral is not easily determined; it depends on the wavelengths of light that the mineral emits (that is, the wavelengths it does not absorb) and which, therefore, reach the eye of the observer. It is interesting to note that colours which appear to be the same can actually consist of various mixtures of coloured lights. Colour analysis of minerals can be carried out with the aid of a spectroscope.

For mineral identification it is important to know the difference between a true coloured mineral and a mineral coloured by impurities. One characteristic, however, the *streak* of a mineral, is constant and easily determined. The streak is found by rubbing the mineral to be identified against a piece of unglazed porcelain (streak plate). It leaves behind a coloured scratch, which, in contrast to the normal colour of a mineral, is constant and typical of each mineral.

Another typical property of a number of minerals is *optical anisotropy*, which can also give valuable details for identification. Anisotropic bodies show different physical properties in different directions. The internal structure of a crystal affects light transmitted through the crystal and the effect varies according to the direction in which the light is transmitted. Amorphous minerals and all minerals which crystallize in the cubic system are isotropic, and have no anisotropic properties.

*Double refraction* is probably the most important aspect of optical anisotropy (as in the Iceland spar type of calcite crystal and some zircons). All optically anisotropic minerals have this property. Anisotropy of the colour of a mineral is called pleochroism, which is the change in quality and intensity of colour due to the crystal's unequal absorption of light vibrating in different planes. The pleochroism of some minerals (e.g. tourmaline) can be observed using ordinary light but usually it is necessary to use polarized light.

The ability of a mineral to refract different colour components of white light at different angles, and therefore break up white light into the colours of the spectrum, is called *dispersion*. This is how rainbow-coloured flashes originate in gemstones. Diamond has an exceptionally high dispersive power.

In the identification of precious stones, special attention should be given to how to distinguish between genuine and synthetic stones. If there is a suspicion that a precious stone is synthetic, it is important to remember that synthetic gems can resemble minerals of the same chemical composition, or minerals of a similar colour but of different chemical composition. This makes it very hard to tell them apart. Distinguishing between genuine and synthetic corundum, especially rubies and sapphires, is rather difficult, for in this instance the natural and the synthetic gems have identical physical and chemical properties.

It is easier to identify synthetic spinel than to identify synthetic corundum because spinels are used for the imitation of other precious stones more often than genuine spinels. They can be identified by finding their optical data. It is harder to distinguish synthetic emeralds, which in the past few years have appeared on the world markets in increasing numbers. However, it is very easy to identify imitation rutile, because natural rutile, unlike the imitation variety, is usually not transparent. Other new synthetic stones are now available for sale, mainly differently coloured imitation garnets.

22 Basalt — Panská skála (Czechoslovakia).

The identification of small mineral grains in rocks is a common task of petrography. It is usually made with the aid of a polarizing microscope.

In conclusion, several other mineral properties are given; they are not commonly used for identification purposes, but are important all the same.

Different minerals have different *solubilities*, according to their chemical composition. If a crystal dissolves only partially, markings called etch-marks appear on its crystal faces. This happens to crystals of many minerals.

Another vital occurrence in some minerals is *radioactivity*, an invisible radiation, caused by decay of atoms of some elements. This radiation affects a photographic plate; if a slab of uranium gangue is placed on the plate, an auto-radiograph will be made. *Magnetism* is another interesting characteristic. There are very few magnetically active minerals, i.e. minerals which attract metal objects (magnetite is such a mineral). However, a large number of minerals are attracted by a magnet, since they usually contain iron.

23 Obsidian — Arrarat (Armenia, USSR); 11 × 8 cm.

24 Perlite — Byšta (Czechoslovakia); 9.5 × 7 cm.

25 Pumice — Mono Lake (California, USA); 11 × 8 cm.

26 Diabase — Prague (Czechoslovakia); 12 × 10 cm.

# THE ORIGIN AND OCCURRENCE
# OF ROCKS AND MINERALS

Mineralogy is not only a descriptive science, but it also explains the origin of minerals. Just as the earth is in a continual state of change, so minerals also develop, grow, alter and decay. The subject of mineralogy studies and explains all these changes. If we wish to understand the laws which govern mineral changes, we must take into consideration the environment in which the particular mineral is found. Minerals which are formed in contact with each other often affect each other's development; this is called paragenesis.

As with chemical elements, there is great diversity in the occurrence of minerals in the earth's crust. Other minerals are practically always present. Quartz is found in the most diverse environments, originating under various geological conditions. On the other hand, the presence of some minerals is very rare indeed. The occurrence of minerals is determined mainly by geological conditions. Many of the deposits are directly dependent on the type of rock in which they originate. Different minerals are found, for instance, in igneous rocks, which are consolidated from magma, from those in the sedimentary or metamorphic rocks. Mineral deposits are divided into classes according to these three types.

The interior of the earth is composed of a molten silicate mass called magma, parts of which gradually solidify and form igneous rocks made up of various minerals. How is it though, that magma, which is more or less homogeneous, can turn by solidification into a number of different minerals?

In *igneous rocks* we usually find minerals which originated at high temperatures. Any homogeneous chemical compound melts at a definite temperature. If we gradually heat a chunk of ice, for instance, the ice does not alter until the temperature of 0 °C is reached. Then the ice suddenly starts to melt. If we carry on heating the ice, its temperature will start to rise only after all the ice has melted. Not until then will the temperature begin to climb. If we gradually cool the water, it will start crystallizing at 0 °C and its temperature will remain constant until all water has turned to ice. Then the temperature will begin to fall again.

Every cooling magmatic mineral solidifies and crystallizes only at a certain temperature, which is called the melting point. The melting point of quartz is 1713 °C, of sulphur 115 °C, of orthoclase 1300 — 1350 °C. When magma comes from deep in the earth to the surface and begins to cool down, its components do not solidify all at once.

Igneous rocks which solidified before they reached the surface of the earth are called *intrusive rocks*. **Granite** (18) is the most common one. Their main mineral constituents are felspar-orthoclase, quartz, and mica, which are formed in grains or flakes. Larger crystals, embedded in fine-grained substantial matter, are only rarely found. Granite is the most abundant rock in the world and is mostly found under the surface of the earth's crust. Granite and its varieties occur in huge masses everywhere. Large blocks of granite are sometimes mined from quarries. Many monumental stones are made of such granite blocks. In Finland, for instance, there is a large granite mass measuring 23,000 km².

Other rocks which have solidified deep below the earth's surface and which are also mainly composed of felspar, quartz and mica, are **pegmatites** (19) (basically a coarse-grained granite, found in veins). Mica is absent in some pegmatites, whereas others may contain dark, instead of light-coloured mica, or they may be of double-mica composition (containing dark and light mica). Graphic granite is a special type of pegmatite. Its main components, felspar and quartz, are intergrown in such a manner that they create a structure which resembles the letters of the Hebrew alphabet. The most important regions where pegmatites occur are in Brazil, in the USA (South Carolina and Maine), in India and in the Urals. Pegmatites are economically important, for their main constituents (felspar, quartz, mica) are in great demand.

Apart from granite and pegmatite, a number of other intrusive rocks are known. They mostly resemble granites but instead of orthoclase felspar they contain soda-lime felspar (plagioclase) and completely lack quartz. **Gabbro** (20), for instance, is composed predominantly of plagioclases and pyroxenes. The largest gabbro massif, measuring 6,100 km², is found in Minnesota (USA); smaller massifs occur in the Harz Mountains (East Germany) and in many places in West Germany.

**Anorthosites** (21) are composed almost entirely of plagioclases. There are immense massifs in Labrador on the east coast of Canada, where 17 anorthosite bodies occupy an area of 130,000 km². Anorthosites from this deposit are composed mainly of labradorite with a beautiful play of colours.

Apart from intrusive rocks, which have been briefly described, and which are composed exclusively of crystalline minerals, there are also magmatic rocks which, more or less, consist of a dark solidified amorphous glassy substance (volcanic glass). These are *extrusive*, or *volcanic rocks* and they differ from intrusive rocks in structure and external appearance. Extrusive rocks solidified on the earth's surface.

There are many types of extrusive rocks in which volcanic glass is dominant. Less frequently found are igneous rocks which are composed solely of this glassy mineral and which resemble dark glass artificially manufactured. They are called **obsidians** (23), and are found mostly in regions which are still volcanically active.

Volcanic glass marked with concentric cracks separating tiny globular particles, is called **perlite** (24).

**Pumice** (25) is a light, frothy volcanic glass whose pores often form the greater part of the rock. It occurs, together with other types of volcanic glass, on the Lipari Islands, in Auvergne (France) and in other places.

Extrusive magmatic rocks have of course developed even outside areas of volcanic activity. Mineralogically the richest rocks are greyish **phonolite** and dark, almost black **basalt** (22) (mostly of Tertiary origin), **melaphyre** (Palaeozoic to Mesozoic origin) and **diabase** (26), a rock of varied age, marked with a conspicuous texture.

In the wake of volcanic activity metamorphic veins and deposits are often formed. Originally igneous rocks may sometimes have their composition altered considerably through the effect of hot (hydrothermal) solutions and gases rising from deep within the earth. These changed rocks are

27 Slate — Austrian Alps; 13 × 9 cm.

28 Limestone layers — Prague (Czechoslovakia).

called metamorphic rocks. Sometimes the effect of these heating processes is to melt out metals and leave them to solidify as veins which it is economically worthwhile to mine.

Tin veins have usually originated from granite, so their minerals are not unlike the granite and pegmatite variety. They consist mainly of cassiterite, fluorite and mica. The largest deposits of this kind are in Bolivia, in the Far East, in the Soviet Union, the USA and in China, in England, Czechoslovakia and East Germany. The gold-bearing veins, whose fissures are composed of quartz and pyrite, or, on rare occasions, of antimonite or arsenopyrite, are poorer mineralogically. They have been found in California and Nevada in the USA, in the Massif Central in France, and in Transylvania in Romania, for example.

The most common type of ore veins are sulphide veins, i.e. veins rich in sulphides. They are divided according to the predominant metal into the following groups: 1) copper ores, 2) silver, lead and zinc ores, 3) antimony ores, 4) mercury ores, 5) iron and manganese ores.

Copper ores occur in veins or beds, which have a stratified or lenticular structure. Their chief mineral is chalcopyrite, and it is often accompanied by pyrite and arsenopyrite. Sometimes the deposit also contains rarer minerals. The most well-known copper ore deposits are at Rio Tinto in Spain, and Butte, Montana in the USA.

Silver, lead and zinc ores are most usually found together in the same deposits, and are therefore called polymetallic ores. The most noted deposits of this type are at Kongsberg in Norway (known for large stringers of pure silver), and Freiberg, which is the centre of the Saxony region of the

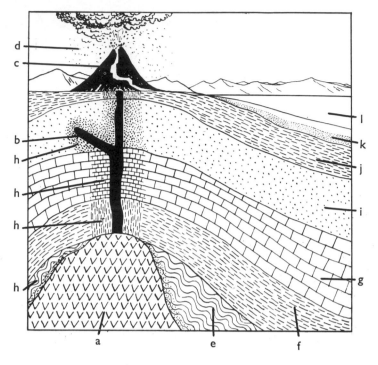

Origin of various types of rocks:
Igneous rocks: a) intrusive rocks, consolidated at great depths, b) vein rocks, c) extrusive rocks, d) pyroclastic volcanic material expelled from a crater
Metamorphic rocks: e) f) g) rocks altered through high pressures and under high temperatures in various depths of the earth's crust, h) rocks formed by contact metamorphism in places of contact with molten eruptive lava
Sedimentary rocks: i) j) k) l) sedimentations of various geological ages

eastern Ore Mountains, where rich deposits are found. Argentite, arsenopyrite, marcasite and fluorite are the most important minerals found here. On the Bohemian side of the Ore Mountains is situated Jáchymov with ores of nickel, cobalt, bismuth, arsenic and uranium. Ore veins of magmatic rocks of the Tertiary Period can be found in Europe in the Carpathian Mountains, for instance. Rich deposits occur especially in Slovakia and Romania, where sulphides, such as sphalerite, galena and pyrite, are abundant. The most important deposits outside Europe are in the USA and Mexico. Hydrothermal ore deposits are particularly interesting. Uraninite is their chief mineral. They belong to the polymetallic ore deposit group. These occurrences are among the most vital sources of uranium. They are found in Canada, Zaïre (Katanga — Shinkolobwe), in Portugal, England and in Fergana in the USSR.

The antimony ore veins, apart from antimonite contain hardly any other mineral. Sometimes though they contain a substantial amount of gold. Mercury ores consist mainly of cinnabar, which is usually accompanied by quartz, carbonates, pyrite, marcasite, antimonite, etc. Iron and manganese veins consist mainly of fibrous haematite and a miscellany of manganese ores.

Mineral deposits often originate through sedimentation. Water streams carry away grains of sediment and sand, or larger particles of disintegrated rock. The faster the water runs, the more grains and particles it washes out and takes away. Gradually, according to the velocity of the current, the fragments are deposited again, the larger particles in the upper course, the smaller and finer ones in the middle and lower courses. Similar transportation and sedimentation occur on land due to the activity of glaciers, winds and accumulation of matter of organic origin. **Limestone** (28) is the most common *sedimentary rock*. Limestone rocks originated in various geological eras and are found almost everywhere in the earth's crust, especially in the form of marine deposits. Often they contain fossils of various marine organisms, such as molluscs, crinoids and corals, and other organisms, which are frequently visible with the naked eye.

In fissures and cracks of limestone beds, crystals and dripstones of calcite, or other minerals, often form. Siliceous nodules composed of flint, chalcedony or opal, formed around sponges, especially from the Mesozoic Era, are frequently present.

Under favourable conditions, pure limestone can develop so-called karst formations, which are due to the solution and leaching of limestone by water. This is the origin of the extensive subterranean cavities, galleries, chimneys and caves, with their magnificent and striking dripstone decor. Well-known karst formations occur in Yugoslavia, Greece, Turkey, the Pyrenees and in Belgium. Bauxite deposits often cover the surface of karst limestones; they are sedimentary rocks composed of a mixture of hydrated aluminium oxides. Bauxite deposits result from the decay and weathering of the insoluble remains of the aluminium-bearing rocks, under tropical or subtropical climatic conditions (Central America, France, Hungary, etc.).

Sedimentary iron ores have been formed by deposition from water on the floors of seas and lakes in various geological ages. Such deposits are abundant and often have a practical significance. Primarily they are haematitic ores, and complex iron silicates. **Clay,** or **pyrite slate,** rich in pyrite and extracted for the manufacture of sulphuric acid in many countries even in the 19th century, belongs to the geologically old sedimentary rocks. They are from the lower Palaeozoic Era and are also mineralogically interesting, because secondary sulphates often originate in their ores.

Some coal-bearing formations are also of interest from the mineralogical point of view, especially the ones belonging to the Permian and Carboniferous system (late Palaeozoic Era). In this case mineral coal was formed by the process (dry distillation) of changing ancient plant material. Petroleum deposits are approximately of the same geological age and probably of the same organic origin. The most important oil fields are concentrated in two large regions; the American region,

29 Sandstone — Hřensko (Czechoslovakia).

with deposits in the USA, Venezuela and Colombia, and the Near East adjoining Europe in the vicinity of the Caspian Sea.

Perhaps the majority of sedimentary minerals originate from deposits from sea water, which contain a great amount of salt, especially sodium chloride. The youngest formations, from the Quaternary Period, are composed mostly of **clay, sand, gravel** and **loess** (sediments formed through the action of the wind).

**Slates** (27) are dense, fine-grained clayey sediments. Similar sedimentary rocks composed mainly of quartz grains are called **sandstones** (29).

No rock remains without changing, whether it was formed deep under the earth's crust, or on the earth's surface. During the past hundreds of millions of years the earth's crust has subsided in places, or has been claimed by the sea, whose bed has gradually become more crowded with magnificent rock formations; this is how rocks which had once originated on the earth's surface have sometimes found themselves in the depths. Exposed to tremendous heat and pressure, they turned into *metamorphic rocks*. There are, however, many regions in nature which are composed jointly of metamorphic rocks and igneous rocks, deeply seated. These are called basement rocks. Metamorphic rocks usually develop there with granite, pegmatite and related rocks.

Serpentinites and amphibolites are important metamorphic rocks low in silicon dioxide. **Serpentinite** usually results from alterations of rocks rich in olivine. The largest and the best known deposits are in the USA, Cuba, New Caledonia and, in Europe, in Greece, Italy, Poland, Czechoslovakia and the Urals. **Amphibolites** are formed through crystallization of igneous rocks that are low in silicon dioxide. Their main components are amphibole and plagioclase. Druses of secondary minerals, composed chiefly of sodium and calcium, often develop on walls of amphibolite rock cavities. This is mineralogically of interest.

**Crystalline limestones,** or **marbles** (30), are rocks of sedimentary origin that have altered through effects of high temperatures deep within the earth into crystalline rocks. They differ from

30 Marble — Supíkovice (Czechoslovakia); 12 × 9 cm.

31 Mica schist with garnets — Solopysky (Czechoslovakia); 11 × 8.5 cm.

32 Chlorite slate — Tirol (Austria); 12 × 9 cm.

33 Quartzite — Prague (Czechoslovakia); 11 × 8 cm.

the compacted sedimentary limestones by the size and texture of their grains. During crystallization, changes sometimes take place in chemical composition. This occurs mainly at the point of contact between the limestone and the intrusive magma. Minerals which then develop (chiefly the silicates) are therefore classed as contact minerals. Graphite beds, which are often directly embedded in the marbles, or form in schists, are closely allied to the crystalline limestone. Minerals of graphite deposits are similar to those in crystalline limestones, but in the whole they are less typically developed (such as pyrites). Among all the world's deposits one of the best known are the famous white marble deposits at Carrara in Italy.

**Gneiss** is the most common of the metamorphic rocks, but is mineralogically poor. It is mainly composed of quartz, felspars, micas, amphiboles and pyroxenes.

**Quartzite** (33) originates through recrystallization of quartz sediments, especially of sandstone.

**Crystalline slates** are metamorphic rocks of schistose texture. They are commonly composed of one sole mineral after which they are named (for instance chlorite slates (32), biotite slates, talc slates, etc.).

Metasomatic deposits of carbonate rock should be included in the metamorphic group. They were formed when the original layers of limestone or dolomite were dissolved and then replaced with less soluble components, such as siderite or magnesite. The best known deposits of this kind are in the eastern Alps.

39

# THE CHEMICAL COMPOSITION OF MINERALS AND MINERAL SYSTEMS

Every mineral has a constant chemical composition, which is expressed by a chemical formula. An absolutely pure mineral hardly ever exists; minerals are usually composed of a mixture of compounds and elements. The mixtures in various minerals are not accidental, for certain elements can exist jointly only within the boundary of the physical laws of chemistry. Galena ($PbS$), for instance, usually contains impurities of silver, which can replace lead, for the crystal structure and physical properties of galena and argentite ($Ag_2S$) are closely allied.

Such blending of chemical compounds of a similar chemical formula is termed as isomorphous. It is quite usual with some minerals for two or more elements to replace each other, under certain conditions. In olivine, for example, iron and magnesium are always present, and they often replace one another. However, the proportions of the constituent elements vary in different olivines. This is why the formula of olivine — magnesium-iron silicate — is expressed in symbols as ($Mg$, $Fe)_2SiO_4$; the elements Mg and Fe can substitute each other. Much can be learned from a chemical formula. It is possible to determine, for instance, the approximate proportion of a metal or other useful element present in a mineral. Every element has a definite atomic weight (ratio of the weight of one atom of the element to one-twelfth of the weight of one atom of carbon): it is sufficient to insert the corresponding numerical values into the chemical formula, then to convert them to percentages. As an example, haematite: ($Fe_2O_3$); atomic weight of Fe = 55.85; atomic weight of $O$ = 16. Molecular weight is the sum total of the atomic weights of the atoms composing a molecule of the substance ($Fe_2O_3$ = 159.70). From this the volume of Fe in a given quantity of $Fe_2O_3$ may be worked out.

Minerals are arranged into systems according to their chemical composition and structural crystallographic properties. In this book they are divided into the following ten classes; 1) elements pages 41—58, 2) sulphides and similar compounds pages 59—87, 3) halides pages 88—93, 4) oxides pages 94—186, 5) carbonates pages 187—200, 6) borates page 201, 7) sulphates and similar compounds pages 202—215, 8) phosphates and similar compounds pages 216—232, 9) silicates pages 233—329, 10) organic compounds (organolites) pages 330—337. The 11th chapter deals with meteorites and tektites (pages 338—344) which, although they do not belong to the mineral system, stand close to rocks and minerals.

The intention of this book is not only to expand the reader's knowledge in mineralogy, but also to capture the mineral kingdom in all its colourful splendour. This is why a specific mineral may not be found in what may be deemed the appropriate place from the strictly mineralogical viewpoint. It was decided not to adhere necessarily to the established systematic order in order that the reader might find the minerals wherever their colourful beauty could be most practically applied and most appreciated. For the primary purpose of this book is to stimulate the reader's interest in the aesthetic beauty of the stones, to study, to collect and explore further the mineral world.

## EXPLANATORY NOTES

The physical and chemical data of the minerals are given in the marginal texts in the following order:
1 Name of mineral,   2 Chemical formula, or chemical composition,
3 Mineral system,   4 Hardness,   5 Specific gravity,   6 Colour,
7 Lustre,   8 Streak.

## ABBREVIATIONS USED:

H. — hardness
Sp. gr. — specific gravity
S. — streak

In the margins the reader will also find characteristic line drawings picturing the most typical or exceptionally interesting crystal forms of described minerals.

40

# Chapter 1 ELEMENTS

Elements, with a few exceptions, are not usually found as minerals. There are 109 known chemical elements, but only 22 of these are found in the native state. In reality there are a few more than 22 minerals in the element group because some elements occur as more than one different mineral — for example, carbon occurs as diamond and graphite.

Pure light metals have so far not been found as natural minerals. They oxidize rapidly and there are no natural conditions that would preserve them in their pure state. In general one can say that the greater the ability of an element to combine with another, the less is the likelihood of its occuring naturally in a pure form. This is the reason, for instance, why in nature pure iron is far rarer than pure gold, though the earth's crust contains more than 4 per cent of iron and only a few millionths of a per cent of gold. Natural alloys of elements, such as amalgams, are classified as elements in the mineral system.

**Copper** (34) was probably the first metal used by man. This is because copper ores are particularly abundant and the working of copper is comparatively easy. Ancient Rome gave copper the name *cuprum* after it had been discovered on Cyprus. Copper occurs chiefly in various ores, usually in compounds with other elements, though pure copper sometimes does occur. Its crystals usually form distorted branching groups. Like most of the minerals which contain this metal, pure copper is generally also conspicuously coloured on the surface. Most frequently it is covered with blue azurite and green malachite. The native metal occurs through the crystallization of hot solutions, or through the conversion of the sulphides of copper in the surface layers of the ore veins (called the cementation zone). The main sources of copper are in Michigan (USA), Australia, Cornwall (Britain) and the Urals in the USSR.

Cu, cubic. H. 2.5–3; Sp. gr. 8.4 to 8.9; copper red; metallic lustre; S. reddish-brown.

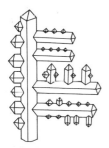

34 Copper — Burra (Australia); 10 cm.

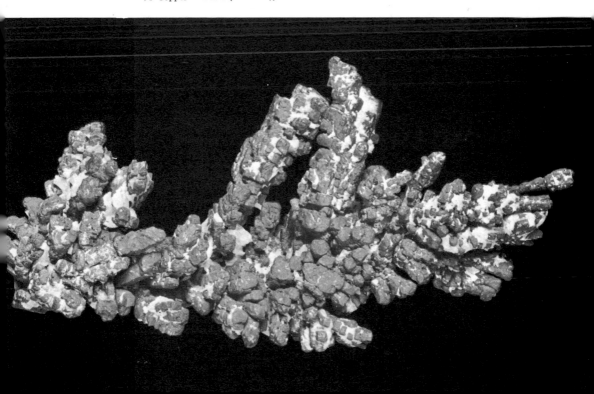

35 Silver — Kongsberg (Norway); 9 cm.

**Silver** (35—37), like copper, occurs not only in compounds, but also as pure metal. It has been used through the ages. Archaeologists have found silver jewelry in the vaults of Chaldean kings from the 4th millennium BC (2). The oldest European silver mines were discovered in Spain. Saxony (Freiberg) and Bohemia were later centres of silver production; it is said that silver was mined there from the 7th century. After the discovery of America, the importation of silver from Mexico and Peru caused a slump in the European mining industry. Thirst for silver was at one time as strong as thirst for gold and, in the 17th century, silver in Japan was actually priced the same as gold. Because of this, the countries rich in silver prospered during the days when large silver deposits were mined. But with the discovery of America came the flow of cheap silver from the New World and prices fell.

Ag, cubic. H. 2.5–3; Sp. gr. 9.6 to 12; silvery-white, yellow on the surface, with brown to black shades; strongly metallic lustre; S. silvery-white.

Pure silver usually occurs as distorted wires, often intertwinned. Other

42

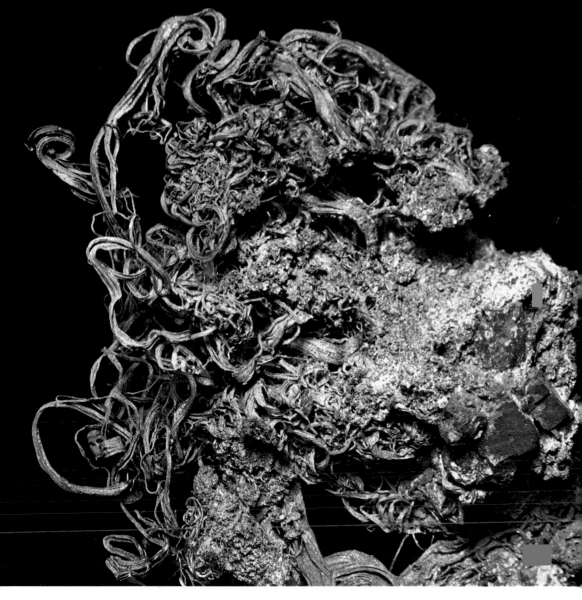

36 Silver — Příbram (Czechoslovakia); 11 × 7.5 cm.

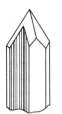

irregular shapes, such as plates, sheets, crusts and coatings, compact lumps, and isometric crystals are more uncommon. Silver usually occurs through the decomposition of silver sulphide ores, for example, of galena and argentite, in the upper layers of the rock (it is called the oxidative and leached zone). The chief European deposits of today are at Kongsberg in Norway; other non-European deposits are in Mexico and North America (Lake Superior, Black Hills, California, Alaska, Arizona, Utah, Nevada and Montana), and at Zmeinogorsk in the USSR. Some deposits, such as those in the vicinity of Schneeberg, Karl-Marx-Stadt (East Germany) and in Peru have yielded substantial silver boulders, some of which weighed as much as several tonnes. In Jáchymov, Bohemia, where in the 16th century large quantities of silver were still being recovered from the rich surface veins, silver wires up to 30 cm in length were found. Similar finds have been made at Kongsberg in Norway. In the Saxony

43

37 Silver — Jáchymov (Czechoslovakia); 11 × 7.5 cm.

deposits, particularly in the famous Himmelsfürst mine at Freiberg, at Schneeberg and others, it was common to find the silver in fine bars, often more than 40 cm long. Today it is hard even to try to vizualize the immense wealth of these deposits. Freshly found pure silver has a magnificent silver colour and lustre. Both, however, quickly vanish when exposed to air; the silver blackens as it becomes coated with a layer of silver sulphide. It is difficult to use pure silver because of its soft texture. Silver alloys containing other metals are therefore manufactured and are used mainly for coinage, and in the jewelry industry, whereas in medicine and the photographic industry silver compounds are generally used.

**Gold** (38—41) is a metal which has always played a major role in history. Throughout the ages the very thought of gold created visions of riches and power. The alchemists therefore tried to manufacture gold artificially, but without success. Archaeological knowledge shows that gold was most probably the first metal man ever knew. He looked for gold in the gravels and sands of streams and centred all his strength and practical knowledge on recovering it. The oldest gold decorative objects which have been preserved come from the early Stone Age. The value of gold must have always been extremely high, for it was used in coinage as early as the 7th century BC.

Au, cubic. H. 2.5–3; Sp. gr. 15.5 to 19.3; golden-yellow to whitish (when with a high admixture of silver); metallic lustre; S. golden-yellow to silver.

During the Middle Ages Bohemia was the richest gold-producing country of Europe. The alluvium of the south Bohemian rivers yielded substantial amounts. Gold was not only found as dust and grain in the alluvium; the fine grains of gold sometimes intruding in the parent rock, especially in quartzite veins, did not escape attention. This is why the miners turned from the original alluvial placers to recovering gold from the excellent gold-bearing quartz veins. Alluvial deposits are today responsible for approximately only one fifth of the total world's production of gold. The recovery of gold is now concentrated on the gold-bearing quartz-pyrite veins, which are present in rocks generally on the circumference of granite massifs.

Gold is usually scattered in rocks so finely that it cannot be seen with the naked eye. Gold forms soft thin plates, wires or grains only rarely and nuggets are even rarer. The most beautiful gold sheets were found in a boulder near Křepice in southern Bohemia. The largest nuggets come from deposits in America and Victoria in Australia. A nugget weighing 85 kg was found in 1869 near Ballarat.

If gold as such is a rarity, what a rare and exceptional find its perfectly bounded octahedral or cubic crystals must be. Gold does not oxidize and, under normal conditions, does not combine with other elements, so it is usually found in nature fairly pure, and only rarely as a compound. It occurs in the original rock either through deposition from hot water or through decomposition of gold-bearing sulphides, mainly pyrite and antimonite, in which traces of gold are often found.

The native alloy of pure gold and silver, which occurs in deposits of pure gold, is called **electrum.** It was thus named because of its pale yellow colour ('electrum' = amber).

Until recently, 50 per cent of the world's gold production came from the gold-bearing conglomerates of South Africa's Transvaal (Witwatersrand and Odendaalsrus) (14). Other major deposits are in America (Mother Lode in California, Butte area of Montana and Lake Superior), Canada, Australia (Broken Hill, NSW and Mount Morgan, Queensland) and the USSR. In Europe, there are deposits in France, Romania and Slovakia. Gold is not only suitable for the manufacture of jewels (3, 4, 7), but also for medicine, coinage and the glass industry. It is important in engineering (contacts for example) and in dentistry it is used in alloys. Gold is perfectly malleable and ductile. It is possible to beat it into such a fine leaf (0.00014 mm thick), that it becomes transparent yellowy-green; from a single gramme of gold a wire of up to 160 metres in length can be obtained. Only metals from the platinum group surpass the density of gold. But gold is very soft and is therefore usually cast with harder metals, such as copper, silver, platinum or nickel.

38  Gold — Roşia Montană (Romania); 10 cm.

39  Gold — Kremnica (Czechoslovakia); actual size of sheet 5 cm.

40  Gold — Křepice (Czechoslovakia); triangular crystals 5 mm.

41  Gold — detail of picture No. 40.

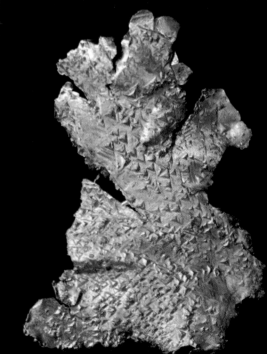

42 Mercury — Istria (Yugoslavia); drops 0.5 mm

**Mercury** (42) was known to the ancient Greeks in the 7th century BC, and they recovered it from the Almaden region of southern Spain. Mercury's scientific symbol, Hg, is derived from the Greek 'hydrargyrum': 'hydor' (water) and 'argyros' (silver). Ancient Romans called mercury 'argentum vivum' giving rise to the old English name 'quicksilver'. Most mercury is extracted from cinnabar ores, in which it occurs as a secondary metal through cinnabar's decomposition. It is also found in its pure form. Pure mercury is an exceptionally interesting mineral. It is the only liquid mineral (other than water). It appears as small drops, usually in small quantities. Mercury was originally used mainly in the recovery of silver and gold from ores and for the manufacture of mirrors. Today its chief use is in medicine, in the manufacture of ammunition, in working precious metals and in the paint industry. Main sources occur in Idria in Yugoslavia and Almaden in Spain.

Hg, liquid, solidifies at a temperature of –38.9 °C and crystallizes in rhombohedral, or hexagonal form. H. 0; Sp. gr. 13.6; tin-white, metallic lustre.

43 Amalgam — Moschellandsberg (Bavaria, Federal Republic of Germany); 7 × 3.5 mm.

**Amalgams** (43) are alloys of mercury and other metals. Mercury has the ability to dissolve many metals and to form new liquid or solid alloys. This is why mercury has always been used and always will be used in mining for precious metals, gold and silver. The rocky gravel containing ores of these metals intermingles with mercury; amalgams develop and through the vaporization of mercury, pure metal results. It is most widely known that silver amalgams also occur as a natural mineral. They appear in the form of tiny marbles or grains or flakes. They always accompany drops of mercury, for they develop jointly through the weathering of cinnabar. The main deposits of silver amalgams are in Bavaria (Moschellandsberg) and Chile. In Colombia even a natural gold amalgam has been found.

Alloys of silver and mercury (Ag, Hg), cubic or hexagonal. H. 0–2.5; Sp. gr. 12–13; silvery-white with metallic lustre; S. grey.

49

44 Platinum — Nizhniy Tagil (Urals, USSR); 5 cm, weight 84.8 g.

**Platinum** (44) is one of the most important and the most valuable, natural pure metals. But there were times when it was cheaper than silver, which it is so alike in appearance and for which platinum had originally been mistaken. Platinum was not discovered until the 16th century in the placer deposits of gold in the alluvium of the River Pinto (today's San Juan) in South American Colombia. It did not catch the attention of scientists until 1748, when a Spanish traveller, Ulloa, told of the discovery of 'silver' which refused to melt. (The melting point of platinum is 1772 °C while that of silver is 962 °C). Platinum usually occurs as grains or thin scales in alluvium, where it has been carried from the magmatic rocks with a small content of silicon dioxide. The world's largest deposits are near Sudbury in Canada and in the Urals. Smaller deposits occur in South Africa and in the USA (Oregon, California, for example), also in the river basin of San Juan near Bogotá in Colombia. Its chief use is in the manufacture of chemicals and jewelry. Platinum consumption is constantly rising. Platinum, and similar metals, such as palladium, osmium and iridium, are extremely dense and are used chiefly in electronic engineering, metallurgy and chemistry.

Pt, cubic. H. 4–4.5; Sp. gr. 14 to 19 (21 when pure); steel-silver, with metallic lustre; S. grey.

**Sulphur** (45) is one of the first elements man ever came across. As it was of volcanic origin and burned with ease, ancient man thought it had supernatural powers. In the past the Romans knew sulphur chiefly from the Sicilian deposits which even today have generous yields. It did not take long before they found various uses for sulphur, especially when they discovered the ease with which it ignites and burns with its intense blue flame, giving off sharp, irritating fumes which bring on tears and coughing. During the 18th century many natural scientists pondered over the cause of sulphur's high burning capacity and they concluded that sulphur must be a mixture of vitriol and a specially inflammable matter which they called 'phlogiston'. That sulphur is an element was discovered only in 1809, when scientific chemistry started developing.

There are many minerals which contain sulphur in nature. They are therefore mined wherever they are concentrated in large deposits. Occasionally sulphur occurs even as an element. Sulphur deposits are usually of volcanic origin. There are other known occurrences caused by hot-water deposition and through the effects of organisms. Sulphur is also abundantly present in all coal pits, where it forms through sublimation during burning. It often develops through the decomposition of some of the sulphur ores in the form of tiny crystals.

, orthorhombic. H. 1.5–2; Sp. r. 2.0–2.1; yellow to brown, with adamantine to greasy lustre; S. colourless to yellow.

45 Sulphur — Caltanisetta (Sicily, Italy); 11 × 9 cm.

Sulphur crystals are usually in pyramidal form, sometimes tabular, usually grouped in druses. They occur in crevices of deposits and sometimes are as much as 10 cm long. They have a strong, diamond lustre, and their colouring is light yellow to honey brown. But the crystals are so fragile that they often crack on handling, because of the warmth of the body. Their weight is low and so is the degree of hardness. Perfectly bounded faces of sulphur crystals are extraordinary. The most magnificent crystals come from Sicilian deposits, especially round the towns of Caltanisetta, Roccalmuta and Grotta. They are in mineral collections all over the world. Sulphur occurs even more frequently in the form of irregular masses or sheets, grain or star-shaped aggregates and impregnations. The aggregates often have a kidney-shaped surface and form stalactites and stalagmites. The sulphur coatings are usually earthy.

The 50 km wide belt between Mount Etna and Agrigento in Sicily is one of the richest sources of sulphur, which, until recently, yielded almost a quarter of the world's sulphur supply. But in the first half of this century mining in Sicily was exceeded by high North American production. Sulphur beds in Sicily are associated with gypsum in limestone layers. They yield sufficient quantities of coarse-grained sulphur, which are extracted by the original, simple method of melting and cleansing in many Sicilian localities. There is, however, an increasing use of modern plants, which make the old-fashioned, primitive ovens called 'calcaroni' obsolete; sulphur itself was used in these as fuel, which was obviously the source of heavy financial losses. Apart from Sicily, other large European sulphur deposits are in Poland in the Tarnobrzeg region.

The finely crystallized sulphur found on some of the Italian volcanoes is of great mineralogical interest, particularly the deposits on Vesuvius, where sulphur occurs in the form of crystallized crusts, or the deposits on the Volcano Island, where sulphur stalactites and stalagmites up to 15 cm in length are found.

America produces 75 per cent of the world's sulphur supplies which amount to 250 million tonnes (Louisiana and Texas with the Boling deposit). The deposits of pure sulphur are of particular importance, because their recovery rests merely on a cleaning process. But the biggest quantity of sulphur for commercial use is not extracted from such deposits, but manufactured by the roasting of pyrites.

Sulphur today is an ever-moving lever of the chemical industry. It would be difficult to name every single process to which it is essential. There are only a few manufacturing processes which could manage without it. Sulphur is mainly used today for the manufacture of sulphur oxides and sulphuric acid. Thus, for instance, production of medicaments, paints and sulphur compounds, for which sulphuric acid is essential, depends entirely on the volume of its manufacture. In powdery form, sulphur is widely used to destroy parasitic fungi and insects (in insecticides), also in wood impregnation, for vulcanizing rubber, for the manufacture of cellulose, and in many preparations of the most varied branches of industry.

**Graphite** (46) has been used for decorating clay vessels. It has often been confused in the past with molybdenite or with lead and antimony ores. In the 18th century it was thought that graphite was a carbon with small amounts of iron. The name graphite comes from the Greek 'grafein' — to write, and was introduced in 1789 by a German mineralogist, A. Werner. A Swedish chemist, J. Berzelius (1779 — 1848) then discovered that graphite is one of the allotropic forms of carbon (another allotropic form is diamond). The great differences in the properties of graphite and diamond lie in the different distribution of atoms in their structure. Carbon usually occurs in foliated, or scaly, sometimes massive form, and develops usually through the metamorphism of carbonaceous material. Graphite deposits are mined in Borrowdale in England, the Adirondack region in the USA, Mexico, the Bohemian Forest of Czechoslovakia and Bavaria, and also in Sri Lanka. Graphite was formerly used for the manufacture of pencils, but today its chief use is for making fire-resistant crucibles, for steel melting and for nuclear technology (in reactors).

Carbon, C, hexagonal or rhombohedral. H. 1; Sp. gr. 2.1–2.3; iron-black to steel-grey, with metallic lustre; S. black.

46 Graphite — Czechoslovakia;
7 × 6 cm.

**Diamond** (47—51) is one of the most valuable precious stones, whose unequalled physical properties place it in a unique position among other minerals. The

Carbon, C, cubic. H. 10; Sp. gr. 3.52; clear, yellow, brown, green, less frequently blue and black, with adamantine lustre.

discovery of diamonds dates back to ancient times. The Roman naturalist, Pliny the Elder (AD 23—79) writes about diamonds, though Romans only knew diamonds of small dimensions. There are many traditional tales about the discoveries and fates of old Indian diamonds, but most of the known large, valuable stones were found only at the beginning of the 17th century. However, the nature of diamond still remained a mystery. The famous physicist, Isaac Newton, stated in 1675 that diamond was combustible. He came to this conclusion because of the diamond's exceptional ability to bend light rays. In 1694, the Italians Averani and Targioni carried out an experiment, during which they burnt the diamond. Sir Humphrey Davy (1778—1829), the British chemist, later proved that diamond actually is carbon.

Since olden days diamonds have been found in the form of grains, or small octahedrons, in alluvial deposits, whence they had been carried from dark, igneous rocks. Efforts to find diamonds in the original rock hardly ever met with success. There are only four known instances of diamonds being found in their parent rock. The oldest one is the famous mine in the Kimberley region of South Africa, where they occur in a decomposed olivine rock, known as kimberlite. Another substantial deposit of diamonds in parent rock is in the Vilyuy river basin in Yakutsk (USSR). In 1961 diamonds in kimberlite were discovered in Sierra Leone and lately also in India, a country already famous for its historic diamond discoveries. The most well-known diamond alluvial deposits are in Zaïre, Minas Gerais (Brazil), in Angola, Tanzania, Ghana and on the west African coast (Guinea, Ivory Coast, Liberia). In America they are found in Arkansas, Ohio, Indiana and Wisconsin.

47 Diamond in kimberlite — Kimberley (Republic of South Africa); 1 × 1 cm.

48 Cut diamonds — Kimberley (Republic of South Africa); actual size of the largest: 23.8 mm in diameter; weight 42.9 carat.

The most famous diamond fields where renowned large stones were found, were in India. Valuable diamonds from local deposits usually became the property of the native Indian princes, rajahs and maharajahs. They were kept in their treasures and passed from generation to generation. Others were placed among the treasures of various temples. These stones did not reach other countries, particularly Europe, till later, mostly as spoils of war. The first news of the large Indian diamonds came into Europe through the efforts of the French researcher Tavernier, who, in 1665, was designated to study Indian treasures. The Republic of South Africa is the best known source of diamonds at the present time, though its output is not the greatest by far. The diamond fields of South Africa were not discovered till 1871. From the time of the discovery until 1920 they produced more diamonds than all the mines of the rest of the world from ancient times.

The history of many of the diamond discoveries is interesting and the fate which pursued the prospectors and the stones themselves is vivid and dramatic. Precious diamonds for jewelry are today usually ground and polished to a brilliant cut of several facets. Diamonds which are very small or imperfect are ground to a rosette cut. Most people are unaware that only about 23 per cent of the diamonds found are suitable for precious gems and that the balance of production is used for industry.

49 Diamond — Jagersfontein near Kimberley (Republic of South Africa); 1.5 × 1.5 cm.

50 Diamonds — Yakutsk (USSR); crystals approximately 7 × 7 mm.

51 Diamond monstrance — (with 6,222 diamonds)
in the Treasury of Loretto, Prague (Czechoslovakia); 89,5 × 70 cm.

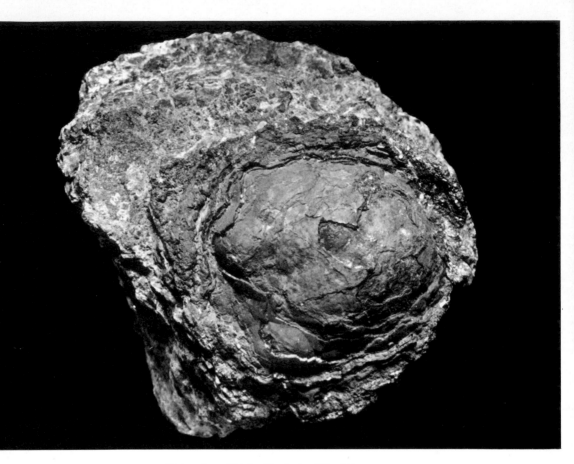

52 Arsenic — Třebsko (Czechoslovakia); 6 × 6 cm.

**Arsenic** (52) and its compounds have been known to be poisonous since ancient times; compounds were obtained from various minerals, mainly from arsenopyrite. As a pure element arsenic is comparatively rare. Its commonest appearance is in round, irregular shell form, with a smooth or granular surface. It usually forms compact masses, fine-grained or reniform (kidney-shaped), rarely imperfect crystals. It is white in colour, but soon turns darker on exposure. It often occurs on silver or cobalt veins. The main deposits are at Freiberg (Karl-Marx-Stadt, East Germany), Jáchymov (Czechoslovakia), Wittichen (Black Forest, West Germany), Japan, British Columbia, Canada and Chile.

As, rhombohedral, H. 3–3.5; Sp. gr. 5.7; tin-white, with dark shades; S. grey.

A natural compound of arsenic and antimony is **allemontite,** named after a location in the French Alps. It commonly forms reniform or cup-shaped structures and is tin-white in colour. Like arsenic, it occurs in ore veins, but is a comparatively rare mineral. Therefore it has no practical significance. Allemontite was discovered in 1822 by the famous French crystallographer, René Just Haüy (1743—1822).

Arsenic today is widely used, especially arsenic which is recovered from its sulphide ores. It is used in making medicaments, paints and pesticide preparations (in agriculture), and for manufacturing glass, impregnating wood, etc.

58

## Chapter 2

SULPHIDES AND SIMILAR COMPOUNDS

This group includes sulphides and similar compounds of arsenic, antimony, tellurium and selenium, which all have similar properties. Sulphides are usually metallic, with only a few exceptions. They used to be divided into blendes, glances and pyrites. Blendes are sulphides of non-metallic appearance, whereas glances and pyrites are metallic; glances are darker and of softer texture, pyrites lighter in colour but harder. This classification is no longer used. Later sulphosalts were recognized and differentiated from the true, simple sulphides. They are more complex compounds, which contain, apart from sulphur and metal, other non-metallic or metallic elements such as iron, cobalt, nickel, bismuth, arsenic, or antimony — which, from the chemical angle, presumably behaved like sulphur. Even this form of classification is not used today. Sulphides are formed mainly from water solutions under very high temperatures.

**Bornite** (53), a conspicuous sulphide of metallic lustre, displays strikingly vivid tarnish colour. It is compact or granular, only occasionally crystallized. It occurs on hydrothermal veins of copper ores, mainly in their lower layers. The largest deposits are in Butte, Montana (USA) and Mexico. Lovely specimens come also from European deposits in Salzburg (Austria) and Vrančice (Bohemia).

Sulphide of iron and copper $Cu_5FeS_4$, cubic. H. 3; Sp. gr. 4.9—5.3; copper red to bronze brown, vividly shaded; metallic lustre; S. black.

**Chalcocite** (53) is more abundant than bornite with which it often occurs. It also more often forms thick-tabular or short-columnar crystals. Finest specimens are found in the Anaconda mine, Butte, Montana. Both chalcocite and bornite are important ores of copper.

Copper sulphide $Cu_2S$, orthorhombic. H. 2.5—3; Sp. gr. 5.7 to 5.8; grey with metallic lustre; S. grey.

53 Chalcocite (light) and bornite (dark) — Vrančice (Czechoslovakia); actual length of vein 12 cm.

54 Tetrahedrite — Capnic (Romania); crystals 8.5 mm.

**Tetrahedrite** (54—55) is thus named because it usually crystallizes in tetrahedral form. It is more often found as coarse-grained, fine-grained or compact masses and its fracture is conchoidal or uneven. Tetrahedrites which contain arsenic are of a lighter colour (grey-white) than those with antimony (almost black).

Complex sulphide of antimony and copper, cubic. H. 3—4.5; Sp. gr. 4.4—5.1; black to grey, metallic lustre, greasy on fracture; S. grey, with semi-metal varieties also red to cherry-red.

'Tetrahedrite' is actually a collective name for a whole group of sulpho-arsenides and sulpho-antimonides, which often intermingle in a most varied way. Because of this, some of its varieties have been named as independent minerals. **Tennantite,** thus named in honour of the British chemist, Smithson Tennant (1761—1814), is mainly composed of arsenic; the Austrian **schwazite** contains a substantial proportion of zinc, and the Saxon **freibergite** has a 30 per cent silver content.

Tetrahedrites occur in veins, usually with other minerals, such as chalcopyrite, bornite (copper and iron sulphide), and other sulphides,

60

55 Tetrahedrite — detail.

especially pyrite, sphalerite, galena, and argentite; sometimes it occurs in cavities with pure wire silver. On rare occasions it is found on contact deposits and in copper slates.

There are large tetrahedrite deposits in the Alps, Romania, near Freiberg in East Germany and in the Slovakian Ore Mountains. Rich druses of beautiful crystals occur mainly in Austria (Brixleg in Tirol), Romania (Boţeşti, Capnic), Great Britain (Cornwall), Algeria (with a content of silver) and the USA (Butte, Montana; Bingham, Utah). Tetrahedrite is often recovered not just as a rich copper ore (30—50 per cent copper), but also as a silver or mercury ore, or as an ore of other elements; this is because it does not have a constant chemical composition. These additional minerals are recovered from tetrahedrite as by-products.

**Sphalerite** (56—57) was given its name from the Greek 'sphaleros' which means deceitful, uncertain. This is because it occurs in such varied colours that even expert miners often failed to recognize it. Even in the 16th century they still mistook it for silver ore; only in the first half of the 18th century was it discovered that sphalerite contains a substantial amount of zinc. It was, however, not used for practical purposes till the 1860s, when a metallurgical method for obtaining zinc from sphalerite was worked out. Until then it used to be thrown away as useless, especially as it was considered an unwelcome impurity in lead ores, which caused difficulties in their metallurgical processing. Sphalerite is usually granular and forms on crevices crystals of perfect cleavage and of strong metallic lustre. It is one of the most common of sulphides. It occurs in ore-veins, usually together with galena or pyrite, and sometimes chalcopyrite (polymetallic ores). The largest sphalerite deposits are in the regions of the upper Mississippi and New York State in the USA, and in Europe on the border of Germany and Belgium. Other important deposits are Bleiberg in Carinthia (Austria), Příbram in Czechoslovakia, Bytom in Polish Silesia, and in mid-Wales, Cornwall, Derbyshire and the Lake District in Britain, where it is known as 'black Jack'. Magnificent specimens of yellow crystals have been found in Spain (Picos de los Europas). That particular sphalerite is ground as a precious stone.

Zinc sulphide, ZnS, cubic. H. 3–4; Sp. gr. 3.5–4.2; brown to black, less frequently yellow, greenish to red, rarely colourless (cleiophane), with greasy to adamantine lustre; S. yellow to brown.

Sphalerite is the main ore of zinc, which is mainly used for metal sheets and covers, being resistant to oxidation. Gallium, indium and germanium are present as impurities in sphalerite and are recovered for the manufacture of transistors.

**Wurtzite** has the same chemical composition as sphalerite, but varies in its internal structure and therefore also in the shape of its crystals (which are hexagonal). Unlike sphalerite it is comparatively rare; usually it is brown to yellowish-brown, reniform or cup-shaped, radially acicular. It is found in association with sphalerite in ore-veins, most usually with galena (polymetallic veins).

56 Sphalerite — Ratibořské Hory (Czechoslovakia); 11 × 7 cm.

57 Sphalerite — Bleiberg (Austria); section of an area of actual size 4 × 3 cm.

58 Millerite — Rapice (Czechoslovakia); needles up to 2 cm long.

**Millerite** (58) is named after the English mineralogist, William Miller (d. 1880). It has been found in cavities in coal seams by miners in the Kladno region of central Bohemia for more than a hundred years. Millerite forms fine hair-like crystals, made up of flexible needles, of brassy yellow colour. These needles sometimes reach a remarkable length (6 cm). The most magnificent examples can today be found in cracks of clayey siderites called pelosiderites, forming spherical structures. The millerite needles can be seen only after the ball-shaped forms are broken up, when they often appear as nicely arranged clusters. The breaking up of the pelosiderite balls is something of a sporting accomplishment. It takes quite an effort and, furthermore, only very few balls contain perfect millerites.

Nickel sulphide, NiS, rhombohedral. H. 3; Sp. gr. 5.3; brassy yellow to green-grey, with metallic lustre; S. black.

Millerite is found in other countries too. There are deposits in the Rhineland of West Germany; in East German Saxony (Freiberg, Johanngeorgenstadt); South Wales and Canada; and in some American states (Wisconsin, for example). Specimens found in limestone cavities near St Louis, Missouri, are particularly magnificent. Though the nickel content is nearly 65 per cent, it is never considered as a nickel ore. This is because nowhere in the world does it occur in workable quantities.

59 Hessite — Boțeşti (Romania); 10 cm.

**Hessite** (59) is one of the few natural tellurides. Its form is usually massive, or the crystals are cube-shaped, often distorted. It sometimes occurs in gold-bearing or silver-bearing quartz veins. Main deposits are Boțeşti in Transylvania (Romania), the Zavadinsk mine in Altai (USSR), in California and in Coquimbo in Chile. Hessite is an uncommon ore of silver.

Silver telluride, $Ag_2Te$, monoclinic — pseudocubic. H. 2.5; Sp. gr. 8.31–8.45; lead to steel-grey, with metallic lustre; S. black.

From the remaining tellurides two others are well worth mentioning: **tellurobismuthite,** telluride of bismuth $Bi_2Te_3$, which occurs in faintly yellow, foliaceous to massive aggregates on gold-bearing quartz veins, and **tetradymite,** sulphotelluride of bismuth $Bi_2Te_2S$. The latter was named from the Greek 'tetradymos' — fourfold, because the crystals found on some deposits were twinned on four sides. It is steel-grey, has a perfect cleavage with a prominent metallic lustre. On rare occasions it is found in association with gold on gold-bearing veins in extrusive rocks and their tuffs (such as the andesite tuffs).

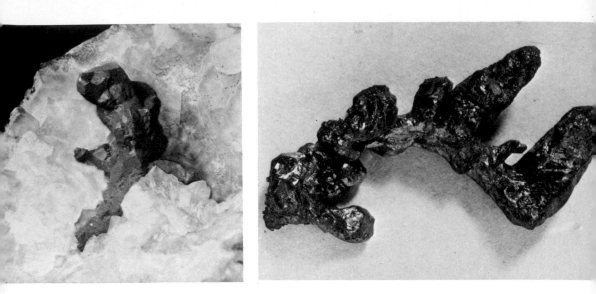

60 Argentite — Jáchymov (Czechoslovakia); 2 cm.

61 Argentite — Ratibořské Hory (Czechoslovakia); 4 × 2 cm.

**Argentite** (60—61), or

Silver sulphide Ag₂S, orthorhombic – pseudocubic. H. 2;
Sp. gr. 7.3; black, with strong
metallic lustre on fresh surfaces, commonly with dark
hues and non-lustrous; S. grey
to black.

silver glance, was the principal silver ore of the Ore Mountains (Erzgebirge) during their reign of fame. Georgius Agricola, 'the father of mineralogy', wrote about argentite in 1529, and described it as a silver ore of
lead colour, occuring in abundance in the mines of Jáchymov. He added
that this mineral resembles galena. In 1598 the same ore was described
by a famous Prague coiner, Lazar Ercker von Schreckenfels in Saxony.
During this era of tremendous mining development, argentite was found
in substantial quantities under the surface of the Ore Mountains deposits
and was responsible for the valuable silver recovered.

Argentite forms compact fillers of veins, or is of wiry shapes which
represent pseudomorphs of pure silver. It is rarer to find it crystallized,
and then the crystals are usually distorted and imperfectly formed. The
pale grey colouring of the fresh surfaces, which have a strong metallic
lustre, resembles galena. The two minerals are easily distinguishable, for
argentite loses the lustre very quickly and darkens, and also can be
scratched and cut with an ordinary knife. In contrast to galena it has no
cleavage, but has a distinct and typically hook-shaped fracture. It is also
noted for its high density. It is found in ore-veins together with other ores
which contain silver, especially the silver-bearing galena. The most
important deposits are at Freiberg and Schneeberg, Karl-Marx-Stadt,
East Germany and Banská Štiavnica in Slovakia. Outside Europe it is
mined in Colorado and Nevada (USA) and near Guanajuato, which lies
northwest of Mexico City. Some of these deposits, such as the ones in
Karl-Marx-Stadt, have yielded boulders of argentite up to 4 kg in weight,
or large interwoven and net-like aggregates. In the Comstock Lode
(Virginia City, Nevada) it occurs compact in large clusters and in abundant impregnations with other noble silver ores. Argentite is, after pure
silver, the richest silver ore with a metal content of more than 87 per cent.

**Chalcopyrite** (62) was well known in the past to miners, for they could not help but notice its vivid surface colouring, which is typical of several copper ores. The medley of rainbow colours, chiefly mauve, blue and reddish, originate from the refraction of light and the splitting up of light rays in the weathered surface layers of the mineral. The true colour of chalcopyrite which has not been through the process of weathering is similar to the colour of pyrite. But it is richer, brassy to golden yellow, and usually bears a faint greenish tint. The identification of these two minerals is therefore simplified when the variability of their hues, crystal form, or the degree of hardness are considered. In contrast to pyrite, chalcopyrite can be scratched with a knife. Chalcopyrite usually appears in veins, in compact or granular form, rarely in crystal form. It occurs most frequently in ore-veins, usually together with other sulphides. The best known deposits are in the Alps (Mitterberg in Salzburg, Austria), Mannsfeld in East Germany (known for the so-called 'copper slates'), Cornwall in Britain and Pennsylvania, Arizona, Montana and Utah in the USA. Chalcopyrite is the most widely spread copper ore found in nature, and it contains nearly 35 per cent copper.

Sulphide of copper and iron $CuFeS_2$, tetragonal. H. 3—4; Sp. gr. 4.1—4.3; brass to gold-yellow, metallic lustre; S. grey-green.

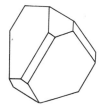

62 Chalcopyrite — Tavistock (Devonshire, England); crystals 2 cm.

63 Molybdenite — Krupka (Czechoslovakia); 4 × 2.5 cm.

**Molybdenite** (63) got its name by mistake, for 'molybdaena' in Greek means graphite, for which it was mistaken. In ancient Greece this name was used not only for graphite and molybdenite, but for other similar minerals, such as some minerals with antimony, which were used for writing implements. In 1778 the Swedish chemist Scheele took it out of this group of minerals with the same name. Molybdenite forms hexagonal, easily flexible scales and flakes, of perfect cleavage, which often closely resemble graphite. It is, however, distinguishable from graphite by its blue to violet hue and metallic lustre. It occurs mainly in granites and pegmatites which are hydrothermally metamorphosed, usually accompanied by cassiterite and wolframite. Nearly 90 per cent of the world's production of molybdenite comes from the deposits in Climax, Colorado. Other important deposits are in Chile and Renfrew in Canada, and in Europe in Yugoslavia, Cornwall and Bohemia. Molybdenite is the most important ore of molybdenum, of which it contains 60 per cent and which is used in the manufacture of steel. It is also used as a solid lubricant.

Molybdenum sulphide $MoS_2$, hexagonal. H. 1–1.5; Sp. gr. 4.7–4.8; lead-grey, metallic lustre; S. grey-green.

**Antimonite** (64—66) was known to the Roman natural scientist, Pliny, who mentions it in his writing as a cosmetic preparation of that time. The silver-grey antimonite used to be finely ground and used for colouring the eyelids. Antimonite occurs in various veins together with other ores, often gold-bearing ores. It forms elongated crystals, which are sometimes of hair-like shape. It is sometimes found in masses. It is well-known for its strikingly beautiful crystals, which were found on the Japanese island of Shikoku, and used as ornaments. The crystals were most popular with the local inhabitants, who used them as flower supports or to make the little fences of the famous Japanese miniature gardens. These unusual crystal creations of the mineral kingdom often also formed an essential part of the interior decor of many dwellings. The strange beauty of the Japanese antimonite vein cavities was indeed most exceptional. The walls of these natural caves were garnished with rich druses and tufts of the most magnificent columnar crystals, as dazzling as polished steel. The mining in Japan has long since ceased and the crystals, which sometimes measured as much as one metre, are today a precious rarity. They can be seen in museum collections throughout the world. The largest deposits of antimonite are in Mexico and in the Kiangsi province of China. The

Antimony trisulphide $Sb_2S_3$, orthorhombic. H. 2–2.5; Sp. gr. 4.63; lead to steel-grey with rainbow shadow, metallic lustre; S. grey.

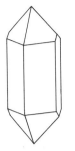

64 Antimonite — Capnic (Romania); crystals up to 4 cm long.

richest European deposits are in the Little Carpathian Mountains and Bohutín near Příbram in Czechoslovakia, in Sardinia and Tuscany (Italy), in Yugoslavia near Fojnica, Krupanj and Kostajnik and in Cornwall in Great Britain. Antimonite is the chief ore of antimony, of which it contains up to 71 per cent. Nearly three-quarters of the production is used for the manufacture of various alloys, which are of great economic value. In printing, for instance, antimony and lead alloys are used for type metal. The presence of antimony increases the hardness of alloys. Bearing metals contain up to 15 per cent antimony, apart from zinc and lead.

65 Antimonite — Shikoku (Japan); 10 × 5 cm.

66 Antimonite — Shikoku (Japan); actual length of section 12 cm.

67 Pyrrhotite — Herja (Romania); 7 × 7 cm.

68 Niccolite — polished plate — Příbram (Czechoslovakia); 13 × 14 cm.

**Pyrrhotite (pyrrhotine)** (67), or magnetopyrite, was given its name because though similar to pyrite and marcasite, it is usually fairly strongly magnetic (when chemically pure and unaccompanied by too many other minerals). It is generally finely granular or massive and only infrequently forms hexagonal crystal plates. Its fracture is yellow-white in colour, but darkens very quickly. It occurs chiefly in some magmatic rocks and in crystalline schists. Either it is found as huge deposits of finely granular pyrrhotite, or only as a fine cover of scattered grain. Another mode of origin is from solutions of ore-veins, where it develops in association with other sulphide ores, or from pyrite from the contact effects of igneous rocks. The largest deposits of pyrrhotite are at Sudbury, Canada and in Wales (Dolgellau) and in southwest England (Botallack near St Just, Cornwall and Bere Alston near Velverton, Devon). Some magnificent crystal specimens have been found in Austria, Romania, France and Brazil. Pyrrhotite is used for the manufacture of sulphur and sulphur compounds.

Iron sulphide FeS, hexagonal. H. 4; Sp. gr. 4.6; yellow-brown to bronze yellow, with metallic lustre; S. black.

Nickel pyrrhotite, known especially from Sudbury, is called **pentlandite**. It is an important ore of nickel.

**Niccolite** (68) was found long ago by Saxon miners while extracting copper ore. They took it also for a copper ore but tried unsuccessfully to obtain copper from it. They named it therefore 'Kupfernickel' ('nickel' means a rogue in German). From this was later derived the name of the element nickel, discovered in niccolite. Niccolite forms compact aggregates on hydrothermal ore-veins. Its main deposits are in La Rioja (Argentina) and, in Europe, in the Saxon Ore Mountains and at Jáchymov (Bohemia). It is an important ore of nickel.

Nickel arsenide NiAs, hexagonal. H. 5.5; Sp. gr. 7.8; greyish red with metallic lustre; S. black.

71

**Pyrite** (69—73) was known in ancient times. Its name is derived from the Greek 'pyr' — fire, because sparks fly from it while it is being broken. In Greece it was considered to have healing powers and was therefore worn as an amulet. According to Pliny it was supposed to stop 'blood decay'. The Incas made mirrors from pyrite. Large polished slabs of pyrite have been found in their graves and this is why it is sometimes called 'the stone of the Incas'. Some practical facts were compiled by natural scientists of the Middle Ages, who based their findings on the experiences of miners. But more detailed examination of pyrite and its distinction from closely resembling minerals is the work of a later age.

Iron disulphide $FeS_2$, cubic. H. 6–6.5; Sp. gr. 4.9–5.2; brassy-yellow with a greyish shadow to golden-yellow, brownish, vividly shaded, with metallic lustre; S. black with a brown hue.

Pyrite occurs in abundance and everyone surely recognizes the small yellow-gold grains and crystals of pyrite in coal, which at first glance look like gold. The black streak of pyrite (in contrast to the yellow streak of gold) soon proves the difference. Metamorphosed and non-metamorphosed sedimentary rocks yield the largest quantities of pyrite. It often forms crystals, in various shapes and combinations, such as cubes

69 Pyrite — Utah (USA); 4 × 4 cm.

70 Pyrite — Khalkidiki (Greece); section 13 × 9 cm.

71 Pyrite — Kutná Hora (Czechoslovakia); 9 × 7 cm.

that are often striated, octahedrons and pyritohedrons. The most perfect crystals come from ore-veins.

Even more commonly, pyrite occurs granular and massive. Its brassy-yellow to golden yellow surface is often vividly tainted, usually blue, green or red. This is because it oxidizes and weathers very quickly — particularly on the surface. It is advisable to take great care of any pyrite specimens in a collection, for as it decomposes it yields sulphuric acid, which often decomposes everything near it. Such alterations of course happen also in nature. Here pyrite oxidizes easily and during this process various sulphates develop, such as gypsum, aluminite (hydrous aluminium sulphate), alum (hydrous aluminium potassium sulphate), and melanterite. Other products of oxidation include the hydrous ferric oxides, limonite and goethite. Pyrite decomposes easily in nitric acid and fuses easily. Its chemical properties make it an unwelcome component of rocks that are used for the exterior ornamentation of buildings, for its presence soon causes rusty, unattractive stains and lines.

Rio Tinto in Spain is an important production centre of pyrite, and so is Japan. Famous crystals come from Lostwithiel (Cornwall) and the

73

72 Pyrite — Rio Marina (Elba, Italy); section 5.5 × 5.5 cm.

73 Pyrite — Gavorrano (Italy); 7.5 × 5 cm.

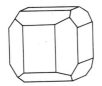

island of Elba, as well as from other Italian deposits (such as Gavorrano), and the Greek deposits of Khalkidiki, Gällivare in Sweden, and from Utah, Wisconsin, Tennessee, New York, Missouri, Montana in the USA. As a raw material, pyrite has many technical uses. Its chief use is for the production of sulphuric acid, sulphates, paints and polishes. It is usually obtained by a roasting process. The by-products can be utilized as low-quality iron ore. Some pyrites carry gold and copper and are therefore used as ores of those metals. The importance of pyrite as raw material for furnishing sulphur has been lessened recently by the mining of pure natural sulphur in newly discovered deposits and by the utilization of industrial gases with a sulphur content derived from coal. An interesting feature of pyrite is that when ground for ornamental purposes, its lustre becomes quite exceptional.

**Marcasite** (74—76) is of the same chemical composition as pyrite. The colouring and a number of other properties are almost identical, so it is not surprising that the name 'pyr' (fire) was, in ancient times, given to marcasite as well as pyrite and that they were both regarded as healing stones right up to the Middle Ages. In 1814 a Frenchman René Just Haüy was the first mineralogist to distinguish successfully between pyrite and marcasite. Today there are various tests to differentiate the minerals.

Iron disulphide, FeS$_2$, orthorhombic. H. 5–6; Sp. gr. 4.6 to 4.9; pale bronze yellow, metallic lustre; S. black with brown shade, paler than that of pyrite.

Marcasite does not have the same internal structure as pyrite. It follows therefore that its crystal form must be different. The crystals are usually small, flat-columnar. They are often twinned in unusual, often very complex units. The names 'spear pyrites' and 'cockscomb pyrites' are given to some of the aggregates. They are basically cyclic 'quins' consisting of short-columnar, elongated crystals. These are of even more complicated form when thin crystals grow in parallel one above the other. Cockscomb pyrite is, in effect, a parallel intergrowth of many twins. The aggregates are often also compact, granular, acicular or radial with a reniform surface, stalactitic aggregates, or just very fine crusts. In coal cracks marcasite occurs even in layers with a reniform surface and is then called 'liver pyrite'. The finely radiating aggregates in coal or the radiating concretions are frequently referred to by miners as 'oranges'. The variability in shape of marcasite is indeed immense. The colour of

74 Marcasite — Lipnice (Czechoslovakia); 10 cm.

75 .Marcasite — Vintířov (Czechoslovakia); 10 cm.

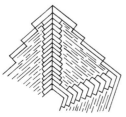

marcasite is almost identical to pyrite. It is also bronze-yellow, but slightly paler, with a greyish-green tint. It is nearly impossible to tell these two minerals apart by their colour. It is best to take into account the difference in the shape of the crystals, which is far more characteristic, and possibly also the lower density of marcasite.

Marcasite and pyrite are both found jointly in coal deposits and in metamorphic and non-metamorphic sedimentary rocks. They occur in concretions embedded in clays and limestones, and less commonly are present in some ore-veins. Even their actual origin has similarities. But marcasite is not so abundant as pyrite. Small quantities of marcasite appear, however, in many places. It is present only in the surface layers of the earth's crust, where, in contrast to pyrite, it is deposited at low temperatures from acid solutions, or develops during the decomposition of plant and animal bodies. If marcasite is placed inside a box, it will turn, after a while, into greyish dust and a stain smelling of sulphuric acid. This happens because marcasite oxidizes even more easily than pyrite. The products of its decomposition are similar. Most frequently it alters into sulphates, particularly gypsum and limonite.

Folkestone and Dover in Britain have the best known marcasite deposits, famous for the chalk spear pyrites. There are large deposits at Freiberg in East Germany and Illinois, Missouri and Wisconsin in the

76 Marcasite — Komořany (Czechoslovakia); 7 × 5.5 cm.

USA. Embedded in coal, marcasite often forms beautiful crystals, either in the cockscomb pyrite form, or the spear pyrite form. Crusts of druses and kidney forms also occur. These are sometimes covered with tiny pyrite crystal cubes. Unfortunately, these marcasites are particularly unstable and are prone to swift decomposition and disintegration.

As marcasite is not as widely distributed in nature as pyrite, it has naturally a lower practical significance, though it can be used for the same purposes. Only a few of the deposits are of practical importance for mining marcasite as a raw material for the manufacture of sulphuric acid.

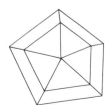

**Proustite** (77) is a member of the so-called 'noble' silver ores. In the past the adjective 'noble' was given to ores with a high content of silver. Proustite was familiar even then because it was so conspicuous with its scarlet colour and high lustre. When exposed to light, however, proustite soon loses the lustre and turns grey, almost black. This happens through the formation of thin layers of metallic silver on the surface (a similar photochemical reaction as on a photographic plate). Proustite is usually granular and sometimes massive, but the crystallized form is less common. It occurs in ore-veins, mainly in deposits of arsenic ores. Some spectacular short-columnar crystals come from Karl-Marx-Stadt (Freiberg and Annaberg-Bucholz), from Jáchymov in Bohemia, from the Mexican state of Chihuahua, from Chile and from Colorado, Nevada, Arizona and Ontario in North America. Proustite is one of the most important silver ores and it contains as much as 65.4 per cent of the metal.

Sulphide of arsenic and silver, $Ag_3AsS_3$, hexagonal. H. 2.5; Sp. gr. 5.57; greyish-red, in light turns grey to black, adamantine lustre; S. lightly scarlet.

**Pyrargyrite** is very like proustite, but instead of arsenic, it contains antimony. When fresh, it is, in contrast to proustite, of a dark red colour, with a metallic to diamond lustre. When exposed to light it acquires the same hues as proustite. It also forms columnar crystals, but more commonly massive, granular or compact aggregates.

77 Proustite — Chañarcillo (Chile); 7 cm.

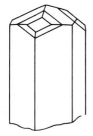

78  Galena — Příbram (Czechoslovakia); 10.5 × 9 cm.

79  Galena — Banská Štiavnica (Czechoslovakia); 9 × 5 cm.

**Galena** (78—79) has been recovered since ancient times and it is mentioned by writers of that period. It is known that lead was obtained from galena in those times, because the process of extracting it was very easy. The Babylonians made lead vases. The Romans manufactured water pipes from lead. After the invention of letterpress printing, lead began to be extensively used as the base of type metal. Galena is a common mineral of ore-veins of various compositions. It occurs in conspicuously granular form, or as cubic crystals. It has a perfect cleavage, and often crumbles readily into small cubes when struck. It is rather heavy. The largest galena deposits are in the valleys of the Missouri and Mississippi in the USA (notably at Joplin, Missouri), and in Europe in Austria, Truro (Cornwall), Derbyshire and Polish Silesia (Bytom). It is also mined in Bleiberg near Villach in Carinthia (Austria) and in Příbram in Czechoslovakia. Galena is the most important ore of lead, of which it contains nearly 87 per cent. It is also the most common ore of silver (up to 0.5 per cent) which is its by-product. Apart from the manufacture of type metal, lead today is principally used for cable covering, accumulator plates and as shielding for radioactive substances. Its compounds are used for the manufacture of paints, particularly the white and red paints, also for lead glass and enamels.

Lead sulphide PbS, cubic. H. 2.5–3; Sp. gr. 7.2–7.6; Grey-black to lead-grey, with a very strong metallic lustre; S. grey.

**Cinnabar** (80) was recovered by the ancient Greeks as early as the 7th century BC, in southern Spain, especially round Almaden. In those days they used it mainly as a red pigment, and called it 'kinnabari' (this gave the mineral its name). But the Romans found it was the source of 'quicksilver', mercury. They also learned how to make an excellent black paint from it. Cinnabar, therefore, became an important mineral. The alchemists of the Middle Ages believed that gold could be recovered from it. Cinnabar forms fine-grained to compact, massive or earthy aggregates and crusts, and sometimes crystals. It occurs most frequently in separate ore-veins. Apart from Almaden, which even today is one of the largest sources of mercury in the world, there are other well-known sources in Tuscany (Italy), Yugoslavia and the USA (especially California). Cinnabar is the chief ore of mercury, of which it contains 86 per cent. Mercury is very important in medicine and in the manufacture of ammunition. Less significant today is its use for working valuable metals, particularly in the extraction of silver and gold from ores and in the manufacture of paints. It is worth noting that the consumption of mercury has not risen during the last few decades, for in some instances, such as in the manufacture of mirrors, other less expensive substances are used.

Mercury sulphide HgS, hexagonal. H. '2–2.5; Sp. gr. 8.1; reddish, with adamantine to metallic lustre; S. scarlet.

80 Cinnabar — Capnic (Romania); 8 × 5 cm.

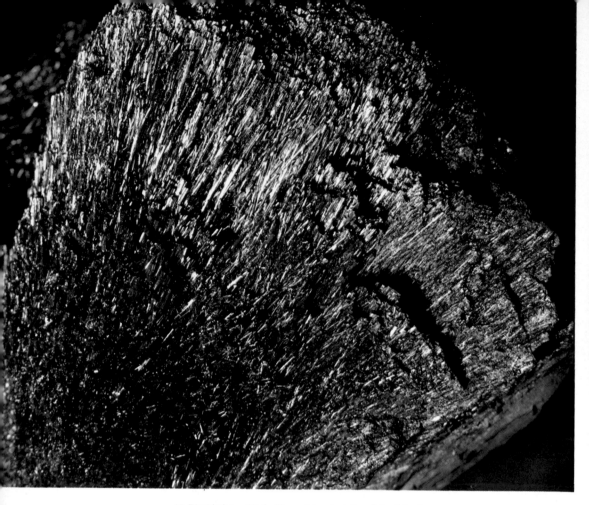

81 Berthierite — Herja (Romania); actual size of detail 9 × 6 cm.

**Berthierite** (81) is so named in honour of P. Berthier (1782—1861), a French chemist, and is not a particularly well-known mineral. It usually forms needle-shaped to compact aggregates which resemble antimonite. The difference appears in berthierite's colour, which is metallic grey with a vivid bronze-brown streak. It occurs in ore-veins, mainly with antimonite. The chief deposits are: Freiberg, Karl-Marx-Stadt district (East Germany), Bohutín near Příbram (Bohemia), Romania and Oruro in Bolivia.

Sulphide of iron and antimony, $FeSb_2S_4$, orthorhombic. H. 2 to 3; Sp. gr. 3.9—4.6; dark steel-grey with bronze shade, and metallic lustre; S. dark brown-grey.

Rather similar chemical composition is shared by **miargyrite**, sulphide of silver and antimony, $Ag_2Sb_2S_4$. This was named from the Greek 'meion' — less, and 'argyrion' — silver, because it contained less silver than other similar silver ores with which it was frequently associated. It occurs in the form of monoclinic minute multi-faced crystals of tabular or other shapes, and in compact or granular aggregates. When fresh, it is dark red, like pyrargyrite, but through weathering it becomes tainted grey to almost black. Miargyrite appears comparatively rarely in hydrothermal veins of silver ores, chiefly in Zacatecas, Mexico, in Braunsbedra, near Merseburg, Halle, East Germany, and in Příbram, Bohemia.

82

82 Boulangerite — Příbram (Czechoslovakia); 8 × 6 cm.

**Boulangerite** (82) is named after a French mining engineer C. L. Boulanger (1810—1849). As it is of the most unusual appearance, it is one of the most interesting and most sought-after minerals. It is usually finely fibrous, while the individual fibres are long, needle-like crystals, grouped into interpenetrant fibrous aggregates. This is why it is often called, together with other rare sulphides, 'feather ore'. Boulangerite occurs in hydrothermal ore-veins together with other sulphides (polymetallic veins). The main deposits are Molières in France, Wolfsberg in the Harz Mountains of West Germany, Příbram in Bohemia, in Polish Silesia and in Nerchinsk in the Urals. Boulangerite is lead ore with 55.2 per cent metal.

Sulphide of lead and antimony, $Pb_5Sb_4S_{11}$, orthorhombic. H. 2.5; Sp. gr. 5.8–6.2; bluish lead-grey to iron black, metallic lustre; S. brownish-grey to brown.

Another mineral from the group of felt-like ores is **jamesonite,** named after the mineralogist Robert Jameson (1774—1854) of Edinburgh, Scotland. It forms monoclinic, finely fibrous dark grey small crystals similar to those of boulangerite, which are also grouped mostly in felt-like aggregates. It is found on hydrothermal beds and in veins of gold-bearing quartz. The principal producers are Arnsberg in Nordrhein-Westfalen, Wolfsberg in the Harz Mountains (West Germany) and Freiberg, Karl-Marx-Stadt (East Germany). Other deposits are at Kasejovice and Příbram in Bohemia, and Zlatá Ida in Slovakia.

83 Arsenopyrite — Salzburg (Austria); 7 × 5 cm.

**Arsenopyrite** (83) was first described in 1546 by Georgius Agricola, who called it 'the poisonous

Sulphide of iron and arsenic, FeAsS, orthorhombic. H. 5.5; Sp. gr. 6.07–6.15; grey to tin-white, often with shadows, metallic lustre; S. dark grey-black.

pyrite'. The ancients had known of its existence, but were unable to discriminate the subtle difference between arsenopyrite and several other pyrites of pale bronze colour. Scientists did not pay too much attention to arsenopyrite until the end of the 18th century, and its chemical composition was not ascertained till 1812. Arsenopyrite is a typical vein mineral, which occurs either with other sulphur ores (polymetallic veins), or individually. In appearance it somewhat resembles pyrite, for it also has yellowish hues. In fresh state its natural colour is tin-white to steel-grey. It is found in compact or crystal form. The largest arsenopyrite deposits are in Sweden. The département of Aude in France, Freiberg in Karl-Marx-Stadt district (East Germany) and some localities in Salzburg, Mexico and Cornwall also have well-known deposits. Though arsenopyrite contains a considerable proportion of arsenic (up to 46 per cent), larger percentages of this element are obtained today from other arsenic ores. The element arsenic and its compounds have been well-known as poisons since ancient times. Today they are used in the manufacture of medicaments, paints and pesticides for agriculture.

**Bournonite** (84—85) was found long ago in the cavities of ore-veins in the Harz Mountains, which are rich in ores. This mineral was also discovered at Příbram in Czechoslovakia, and in Cornwall in Great Britain where it occurred in fairly large quantities in the region near Endellion, and was recovered as a good antimony ore. The mineral was aptly named according to its place of discovery — endellionite. The new name bournonite was given in honour of Bournon, a French crystallographer and mineralogist (1751 to 1825), who described its chemical composition. Bournonite deposits also occur in several places in Carinthia. The crystals of bournonite are mostly tabular with square contours. Sometimes they are intergrown and form interpenetrant twins. The occurrence of low-columnar crystals is less common. Bournonite is also found fairly frequently in the form of well developed crystals in ore-veins, or it is massive, granular to compact, recognizable by its strong lustre. It is comparatively fragile. Bournonite is now recovered primarily as lead and copper ore, for it contains up to 42.6 per cent lead and 13 per cent copper.

Sulphide of lead, copper and antimony, $PbCuSbS_3$, ortho-rhombic. H. 2.5–3; Sp. gr. 5.7 to 5.9; dark grey, metallic lustre; S. steel-grey to blackish.

84 Bournonite — Wolfsberg (German Democratic Republic); crystal 6 × 3.5 cm.

85 Bournonite — Příbram (Czechoslovakia); crystal 2.5 × 1.5 cm.

86 Realgar — Baia Sprie (Romania); large crystal 6 mm.

**Realgar** (86) is one of the natural arsenic compounds, which were once used as paints. It was sold under various names — arsenic ruby, arsenic blende, or arsenic glass. Realgar originates through the decomposition of arsenic minerals, chiefly arsenopyrite, in ore-veins. It forms grains or crystals. The most beautiful specimens come from Allchar in Greece. There it occurs jointly with orpiment in rich deposits of arsenic ores. In a mixture of these two minerals one may occasionally find a bunch of slender columns of antimonite not yet decomposed — a primary antimony ore in this deposit. The specimens from Allchar are one of the most colourful examples of various mineralogical groupings, for the contrast of gold-yellow, blood-red and deep steel-blue-grey is extremely striking. Similar deposits are in Capnic and Baia Sprie in Romania. Getchel mine in Humboldt County, Nevada, USA, is now the principal non-European deposit. Large columnar crystals of realgar are found there. The primary use of realgar today is in pyrotechnics for the preparation of the so-called 'White Flames of Greece', fireworks and in the tanning industry. As an arsenic ore it contains up to 70 per cent arsenic.

Arsenic sulphide, AsS, monoclinic. H. 1.5; Sp.gr. 3.5–3.6; orange-red to dark orange, greasy, vitreous to adamantine lustre; S. slightly paler than surface colouring.

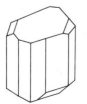

**Orpiment** (87—88), whose name approximately means 'gold pigment', has been used for painting. Once it was exported from the Orient to Europe in the form of large leaves up to 10 cm long, under the name of 'Turkish orpiment'. It came from southeast Turkey near the Iraq border. The natural deposits of orpiment are not sufficient to meet the substantial demand for the manufacture of paints, so it is also produced artificially under the name of 'king's yellow'. In some deposits orpiment accompanies other arsenic ores. It forms through their oxidation and covers them in a yellow crust. Orpiment forms in cavities, compact, cup- or kidney-shaped to round nodules. Crystals are rare. Main deposits: Capnic and Baia Sprie in Romania, Hakâri in Turkey, Nevada and Yellowstone Park in the USA.

Arsenic trisulphide, $As_2S_3$, monoclinic – pseudorhombic. H. 1.5–2; Sp. gr. 3.48; golden-yellow, brownish, with greasy to pearly lustre; S. citrous yellow.

The most perfectly formed crystals of orpiment, hidden in the hollows of arsenic ores, occur fairly abundantly in association with antimonite in the rich deposits near Allchar, northwest of Salonica (Greek Macedonia). Magnificent reniform specimens are found there, composed of fine fibres, frequently intergrown with the blood-red grains of the similar mineral, realgar. Orpiment found at Allchar has a particularly finely layered structure and occurs in substantial quantities.

Orpiment is used as an arsenic ore and contains up to 61 per cent arsenic.

87  Orpiment with realgar — Baia Sprie (Romania); 10 × 8 cm.

88  Orpiment — Allchar (Macedonia, Greece); 10 cm.

# Chapter 3    HALIDES

Minerals that are compounds of the so-called halogen elements, i.e. chlorine, fluorine, bromine and iodine, are called halides. Out of these four elements, the characteristics of fluorine are basically different from those of the other three. Minerals containing fluorine occur therefore in completely different mineral parageneses than minerals with a content of chlorine, bromine and iodine. The most significant occurrences of fluorine minerals are concentrated in granite rocks. The melanochroid magmas and the products of their exhalations are poorer in fluorine but richer in chlorine. This is why active volcanoes with a predominance of melanochroid lavas show only traces of fluorine, but have a large quantity of chlorine. Chlorine, bromine and iodine are, however, amassed in inorganic sediments, particularly in the sedimentations from sea waters. Halides are rather poorly represented in the mineral kingdom in comparison with other groups, as far as the abundance of the individual minerals is concerned. Only fluorite and common salt are plentiful.

**Common rock salt (halite)** (89—90) was, in ancient times, an important object of trade. It was, without doubt, man's most essential mineral and today still remains one of his primary needs. Common salt is mentioned in the works of many ancient natural historians and philosophers. Pliny the Elder (AD 23—79), who was killed during the eruption of Vesuvius, listed various methods for recovering common salt from sea water or salt springs by evaporation. On the Mediterranean coast there were deposits of rock salt in Spain in the surroundings of Barcelona (Cardona). Rock salt was extracted from vast mines in India during the reign of Alexander the Great.

Sodium chloride, NaCl, cubic. H.2; Sp. gr. 2.1–2.2; colourless, grey, rarely reddish or blue, vitreous to dull lustre; S. colourless.

Mining in the Austrian 'Salt Chamber' goes back even further, for archaeological excavations have proved that rock salt was mined there as long ago as the 15th century BC. The remains of ancient mines have been discovered near Hallstatt and, apart from the underground tunnels and passages, spades, ladders and other mining equipment were found, all well preserved in salt and dating back to the Old Iron Age. Other ancient deposits of common rock salt in Europe are in the region of Wieliczka and Bochnia in Poland, where salt occurs in vast quantities with gypsum in salt clay.

It is quite obvious from all the information extant, that man has realized the importance of salt deposits right through the ages. However, its exact chemical composition was formulated by Sir Humphrey Davy, the British chemist and physicist, only in 1810.

The main deposits of rock salt are in the sea. In all the oceans and seas of the world there must be approximately 20 million cubic kilometres of salt; for sea water contains up to 3.7 per cent of rock salt and other salts. This is why along sea coasts salt has been recovered through the ages by simple evaporation of salt water in flat tanks. This method of extraction is particularly common in countries round the Mediterranean and Black Seas.

Deposits of rock salt, in solid form of course, are also found inland, in most geological formations. These were formed millions of years ago by the drying up of salt water in sea basins under hot and arid weather conditions. Salt gradually crystallized from water, thus forming the originally horizontal salt deposits, which subsequently often furrowed. The strongly plastic layers of rock salt thus formed enormous salt beds, which today are the chief sites for salt mining. The stratified deposits sometimes have a capacity of several hundred cubic metres, and the salt is often associated with gypsum and anhydrite, or even with potassium salts.

Common rock salt forms granular to compact masses, and in fissures often occurs in cubic crystals. They have a perfect cleavage and so split with one blow of a hammer into vast numbers of minute crystal cubes. In some coal mines salt stalactites and stalagmites occur; from the mineralogical view they are of great interest. When chemically pure, salt is transparent; the grey coloration is caused by clay, red by haematite, brown by bituminous matter, and blue by the presence of sodium metal. Rock salt commonly contains impurities such as magnesium and calcium chlorides. It is perfectly soluble in water.

The main producers of salt are the USA (Louisiana), Great Britain (Cheshire) and West Germany. The most notable deposits are the two already mentioned in Austria and Poland. Some very beautiful and interesting crystals form on shores or surfaces of evaporating dry lakes (15), such as the Dead Sea, the Great Salt Lake of Utah and the salt lakes of North Africa. Crystal formations also occur on prairie land.

The most beautiful crystals of rock salt are today found in the Siberian salt mines near Iletskaya Zashchita and also in several localities in the United States, such as the salt Lake Searles, San Bernardino County in California. Crystals found in clusters in these deposits commonly reach a length of 2 to 8 cm, and individual crystals can be up to 30 cm long.

Salt is vital to human life. Each of us consumes yearly approximately 7.5 kg. A vast amount of salt is used today in the chemical industry, for

89 Rock salt (halite) — Hallstatt (Austria); 10 × 8 cm.

90 Rock salt (halite) — Wieliczka (Poland); actual size of detail 30 cm.

which salt is an essential raw material, especially for the production of sodium compounds, chlorine and chlorine compounds. The large clear crystals are used in optics.

The chief method of producing salt is by extraction from rock salt deposits, whereas the ancient, simple method of extraction from salt water is very much on the decline. The annual production of salt is roughly 35 million tonnes.

**Sylvine,** potassium chloride, $KCl$, crystallizes similarly to rock salt, often with admixtures of sodium and magnesium chloride. It forms cubic, perfectly cleavable crystals or granular, occasionally also needle-shaped aggregates, which are also highly cleavable. It is colourless to whitish, also has shades of blue, or reddish to red and occurs in salt deposits, particularly at Stassfurt (East Germany) and Kalusz in Poland. Sylvine is the most important potassium salt and has great technical significance, chiefly in the preparation of potash fertilizers (such as kainite) and in the manufacture of potash soaps. Metallic potassium is used as a catalyst in the manufacture of synthetic rubber.

**Fluorite** (91—93) is a mineral which has always attracted attention because of the beauty of its crystals and variety of its colours. It caught the eye of the Ore Mountains miners, who found it while mining for tin, and who called it 'the ore flower', and strangely enough, British miners in the Middle Ages also gave fluorite such a poetic name.

Calcium fluoride, $CaF_2$, cubic. H. 4; Sp. gr. 3.1–3.2; variously coloured, most commonly violet, green, yellow, less frequently colourless, with vitreous lustre; S. colourless.

They were not the first admirers of the magnificent colours and shapes of fluorite, for the ancient Greeks used it as an ornamental stone. They turned it in their workshops into beautiful multicoloured vessels, which later became famous as 'murrhine vases' (from the Latin 'murrha' = fluorite). The natural fluorite was imported from Parthia (today's Iran). During the time of the Roman Empire, these vases were highly valued. As there was a shortage of this rare mineral, glass imitations were produced by a special process. These are some of the first known imitations of precious gems.

Beautiful crystal specimens of fluorite occur fairly abundantly, often with fluorescent colouring. The crystals have a perfect cleavage. They can form compact, granular or wiry aggregates. In 1824, a German mineralogist Friedrich Mohs (1773—1839) discovered that fluorite exhibited a striking phenomenon which was later named fluorescence, after fluorite. This characteristic is particularly noticeable in some of the transparent fluorite varieties. When exposed to light of one colour they immediately emit light of a different colour. For example, when ultraviolet light strikes fluorite it emits violet light. (The term 'fluorescence' is now also applied to many different types of luminiscence.) The colour changes sometimes even on individual crystals or aggregates. Apart from occurring in deposits of tin ore, fluorite is often found in pegmatites and ore-veins where it developed under high temperatures through crystallization from hot solutions in association with volatile compounds escaping from magma. Fluorite of a sedimentary origin occurs less commonly and is found, for instance, in cavities of limestone or in coal. The richest deposits of commercially important fluorite are of this category; they occur in the limestone beds of Kentucky and Illinois, USA.

The greatest deposits of fluorite are in the USA. The best known ones in Europe are Freiberg, Karl-Marx-Stadt, East Germany; Weardale near Durham, Cornwall, the Lake District and Derbyshire (a variety called 'blue John') in England. There are also deposits in France, Italy and the USSR. It is the most abundant natural compound, and is used as a source of gaseous fluorine which was named after this mineral. It is also important as a flux for assisting fusion. The colourless pieces are ground and used for lenses which transmit ultraviolet rays.

There is still much to be learnt about the gaseous element fluorine. The properties of its compounds are, on the whole, quite exceptional. One can certainly say that it is sure to occupy a place of importance in the technical world of the future. Fluorine and some of its compounds are poisonous, so they are used for disinfectant purposes. Some compounds of fluorine are used as refrigerants. Other compounds are used for the manufacture of plastic materials, including non-stick coatings for cooking utensils. Fluorine is an aid in separating uranium from its isotopes and in rocket technology. Fluorite is also cut and polished as a semiprecious stone.

91 Fluorite — Val Sarentina (Italy); crystals 5—15 mm.

92 Fluorite — Durham (England); crystals 2—3 cm.

93 Cut fluorites — England; actual size of the largest 22.5 × 18.1 mm (24.36 carat).

# Chapter 4    OXIDES

Oxides are compounds of oxygen with metallic or non-metallic elements. They are divided into anhydrous (such as quartz, cassiterite) and hydrous (opal, goethite) oxides. The spinel group, for instance, spinel and magnetite (binary oxides with bivalent or trivalent elements), are also classified as oxides though in some old books they were classified as an independent group. The individual members of the oxide group are often isomorphous and isostructural. Another smaller mineral group, once classed as the manganate group, now comes under the heading of oxides. They are also binary oxides but unlike spinel they consist of tetravalent manganese and some bivalent element (usually also manganese). Psilomelane is one such mineral. Under the new system wolframite is also classified as an oxide; it used to be classified as a wolframate. The origin and the occurrence of oxides is extremely varied.

**Cuprite** (94) was first described by Agricola in 1546. He mentions a red-coloured mineral which occurs in the surface layers of some copper ore-veins. This mineral was more accurately described by a French mineralogist Romé de l'Isle in 1783. He referred to the renowned ore deposits in Cornwall in Great Britain, and Timiş in Romania. W. Haidinger, a Viennese mineralogist, gave the mineral the name cuprite in 1845 (from the Latin 'cuprum' — copper).

Copper oxide, $Cu_2O$, cubic. H. 3.5–4; Sp. gr. 5.8–6.2; reddish-black, with adamantine to submetallic lustre; S. pale red.

Cuprite, which is commonly called red copper ore, can be compact, finely granular, massive or earthy; occasionally it forms small crystals. When needle-like it is called chalcotrichite (from the Greek 'chalkos' — copper and 'thrix' — hair). The natural colour is dark red, with a diamond lustre, but has a metallic grey streak. Earthy cuprite, which is usually mixed with limonite, is red to brown-red in colour (the brick ore).

Cuprite is a by-product of incomplete oxidation of copper or its sulphurous ores. This is why it usually occurs only in the upper layers of veins, chiefly in association with pure copper. Its beautiful octahedral crystals of deep cochineal colour are found in cavities of ore-veins. In many deposits such octahedrons are accompanied by chalcotrichite, whose needles in reality are the rather elongated tiny cubes of cuprite, or by other secondary copper minerals. As a result of weathering, cuprite changes into secondary copper carbonates (malachite and azurite). Main deposits: Romania, West Germany (Siegen in Nordrhein-Westfalen), Yugoslavia (Sinjako in Bosna), France (Chessy near Lyon, where octahedral crystals of up to 3 cm occur), the Urals, Arizona (Jerome, Clifton and Bisbee), Australia (Broken Hill in New South Wales), Bolivia, and England (Liskeard and Gwennap in Cornwall).

Next to pure copper, cuprite is the richest copper ore with up to 88.8 per cent copper. The transparent cuprite crystals are very valuable, but occur very rarely.

Another anhydrous oxide of copper is known as **tenorite** (named in honour of M. Tenore (1780—1861), a botanist from Naples, or **melaconite** (from the Greek 'melas' — black and 'konis" — dust), or **copper black.** This is a copper oxide, CuO, and develops chiefly in black earthy aggregates on the surface of copper ores. Its monoclinic pseudohexagonal tabular crystals are a rare find. These are, in contrast to the powdery varieties, steel-black with a metallic lustre. The best known European occurrences are on Mount Vesuvius in Italy, near Siegen (Nordrhein-Westfalen) and Daaden (Rheinland-Pfalz), West Germany, in Jáchymov, Bohemia and in a few localities in Romania. The largest deposits are in the United States, particularly in the vicinity of Bisbee, Arizona, where they form imposing stratifications similar to asphalt, and by Lake Superior.

94 Cuprite
Liskeard (Cornwall, England); 9 × 6 cm.

95 Spinel (pleonaste) — New York (USA); crystal 11 mm.

**Spinel** (95—96) apparently received its name from the Greek 'spinther' — a spark, and this was probably on account of the sparkling colours of its gemstone varieties.

Magnesium aluminium oxide, $MgAl_2O_4$, cubic. H. 8; Sp. gr. 3.5; variously coloured, most commonly scarlet (spinel ruby); yellowish, pink, bluish-pink (rubicelle), vitreous lustre; S. white.

Spinel has been sought after and used for ornamental purposes since days of old, though it is less valuable than the ruby it resembles. Spinel's crystals are not usually large; bigger pieces are therefore a rarity. A few are in the possession of the British Natural History Museum in London. A spinel which weighs 400 carats is among the Soviet collection of state treasures, and a large scarlet spinel is among the English crown jewels.

Though various chemical impurities, particularly bivalent and trivalent iron and chromium, do not disrupt the homogeneity of spinel's crystals, they influence its colouring. For example, red colour is caused by the presence of chromium, reddish-brown colour by trivalent iron, and bluish hues by bivalent iron. Mixtures of elements bring varied mixtures of colour. Chemically pure spinel, produced artificially, is colourless. Both bivalent and trivalent iron are present in substantial quantities in the black **pleonaste**, which frequently contains also chromium. Pleonaste is sometimes considered to be a different mineral. Many gem varieties of spinel are known by their common names used by jewellers, such as the blood-red **ruby spinel,** or the pink-red **balas-ruby,** or the blue-red **almandine spinel** and the hyacinth-red to straw-yellow **rubicelle.**

Spinel occurs in crystal form, but is more often finely granular. It is found in igneous rocks rich in aluminium oxide, and in dolomitic limestones (limestones rich in magnesium), but chiefly in alluvial deposits. The main alluvials are in southwest Sri Lanka near Ratnapura and in north Burma; it is also found at Aker in Södermanland (Sweden), in the United States in Andover, Hoboken in New Jersey and Amity in the New York State, and at Zlatoust in the Urals (USSR). Some of the beautifully coloured transparent spinel varieties are used as gemstones. Spinels are also manufactured synthetically and they provide a colourful and wide range of imitations of the natural mineral.

96 Cut and polished spinels from Sri Lanka; actual size of largest 13.8 × 11.7 mm, weight 9.77 carat.

97 Magnetite — Alp Lercheltini (Switzerland); crystal 15 × 10 cm.

98 Magnetite — New Jersey (USA); 8 × 7 cm.

**Magnetite** (97—98) is a mineral which justly deserved all the attention bestowed upon it by the ancient natural historians and philosophers, who were impressed by its magnetism. Pliny the Elder mentions magnetite; he writes about a hill near the river Indus, entirely made of stone and attracting iron. According to the fable, the stones were first discovered on this hill by a shepherd named Magnés, when he noted that the nails of his boots and the iron ferrule of his staff adhered suddenly to the ground. It is said that magnetism — as a physical phenomenon — was named after this legendary figure (another, more probable version ascribes the name magnetite to the locality of its discovery — Magnesia in Macedonia). Even earlier than this, in the 11th century BC, the Chinese knew about magnetism. Magnetite forms granular to compact masses, and occasionally forms octahedral crystals. It is strongly magnetic. Most common deposits are in the 'skarn' ores of igneous and metamorphic rocks. Magnetite, particularly in skarns, is associated with silicates rich in iron — such as garnet — andradite, pyroxene, hedenbergite, various amphiboles and epidote. The most important deposits are the vast mountain formations in northern Sweden, Norway, and the United States. Magnetite is the most valuable ore of iron and contains up to 72 per cent metal.

Iron tetraoxide $Fe_3O_4$, cubic. H. 5.5; Sp. gr. 5.2; black with a bluish shadow and metallic lustre; S. black.

98

**Chromite** (99) at first glance is indistinguishable from magnetite, but it is only slightly magnetic and has a brown streak (magnetite has a black streak). Commonly granular, it often forms fine- to coarse-grained pockets in olivine rocks, where it occurs as an original constituent; also in serpentines, which formed through their decomposition. Rhodesia, South Africa, the Philippines, Turkey and the USSR share in the world's production of chromite. Substantial amounts occur also in the USA and Cuba. Chromite is practically the sole chromium ore, containing up to 46 per cent chromium, which has the property to harden iron. Nearly half of the total production of chromium is used for the manufacture of high-quality chrome steel.

Oxide of iron and chromium, with admixture of magnesium, $(Fe, Mg)Cr_2O_4$, cubic. H. 5.5; Sp. gr. 4.5–4.8; brownish black, with metallic lustre; S. brown.

Another mineral in the group of binary cubic oxides is **franklinite,** an oxide of iron and zinc with the admixture of bivalent manganese and iron, and trivalent manganese $(Zn, Mn, Fe) (Fe, Mn)_2O_4$. There are world-famous deposits at Franklin and Ogdensburg, New Jersey, USA. Franklinite forms black octahedra, often rounded at the edges, which are embedded in white calcite and accompanied by numerous other interesting minerals.

99 Chromite — Klyuchi (USSR); 9 × 6 cm.

100 Chrysoberyl — Maršíkov (Czechoslovakia); crystals up to 15 mm.

**Chrysoberyl** (100 — 102) is not a widely known precious stone, and it is classed as one of the most valuable minerals. It occurs in pegmatites and in metamorphic rocks (gneiss and mica-schists), where it crystallizes in finely tabular and columnar form. After diamond and corundum it takes the third place as the hardest mineral.

Beryllium aluminium oxide, $Al_2BeO_4$, orthorhombic. H. 8.5; Sp. gr. 3.7; yellow-green to yellow, alexandrite (green in daylight, violet-red in artificial light), with vitreous lustre; S. colourless.

Of all the colourful and transparent varieties of this mineral, **alexandrite** (102) is the most valuable one — named in honour of the Russian Tsar Alexander II. This stone is unusual for being vividly green in daylight, but mauve under artificial light. Because of this it is said that alexandrite is an emerald by day and an amethyst by night.

The chief chrysoberyl deposits are in Sri Lanka and Brazil. Recently it has also been discovered in several places in Africa and in northeast USA (Maine, Connecticut and New York). Beautiful, light yellow,

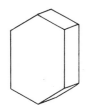

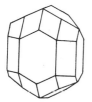

perfectly transparent crystals are found for example by Lake Alaotra in Madagascar; extremely dark alexandrite crystals measuring up to 2 cm occur in south Rhodesia. The best known European deposit is Maršíkov in Moravia. Crystals from Maršíkov are a magnificent yellow, but too flat to be cut and polished. In this locality they occur embedded in aplitic granite rich in muscovite in association with beryl and garnet. The alexandrite deposits in the valley of the Tokovaya river in the Urals, found in 1833, are world famous. Very recently alexandrite has been found at deposits of other chrysoberyl varieties in Sri Lanka and in Brazil. Large chrysoberyl crystals are rare. The largest crystal was discovered in the alluvials of Sri Lanka. The specimen weighed 19 grammes and measured 23 × 17 mm. Chrysoberyl pebbles with a silky lustre are sometimes found in fairly large pieces. After being cut to a lenticular shape, they exhibit a prominent, fluctuating lustre. They occur chiefly in Sri Lanka and Brazil. The loveliest and largest group of alexandrite crystals was found in the Urals. It consists of a druse of 22 crystals, weighs 5 kg and measures 25 × 15 cm. This exquisite specimen can be found in the Mineralogical Museum of the Academy of Sciences in Moscow. An even larger individual alexandrite crystal was discovered in Sri Lanka. Cut and polished, it weighs 66 carats and is in the collection of the Smithsonian Institution in Washington. Chrysoberyl crystals measuring up to 18 cm have been recently discovered in pegmatites near Golden, Jefferson County, Colorado, USA. The price of chrysoberyls, particularly alexandrites, is very high, especially when dealing with perfectly transparent stones. A first class alexandrite weighing 10 carats would fetch as much as $ 25,000.

101 Cut and polished chrysoberyl — Sri Lanka; 18.8 × 17.4 mm; weight 27.81 carat.

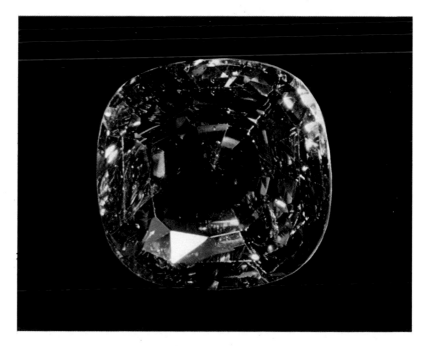

102 Chrysoberyl — alexandrite — Tokovaya river (Urals, USSR); 5×4 cm.

103 Valentinite — Příbram (Czechoslovakia); 6 × 5 cm.

**Valentinite** (103) was named in the middle of the 19th century in honour of Basilius Valentinus,
a writer on alchemy, whose identity is rather a problem. There are
considerable differences of opinion as to when exactly he was supposed
to have lived (16th or 17th centuries). But he is the supposed author of
the first book to give a detailed description of antimony and its compounds. From the contents of the book it is also obvious that Valentinus
was familiar with the synthetic preparation of antimony trioxide, which
was called 'the antimony flower'. Valentinite was first discovered near
Allemont in France towards the end of the 18th century; the first description of its occurrence in the region of Příbram in Bohemia comes
roughly from the same time. This particular locality at one time produced
the very best crystals of this mineral. The largest crystals found there
measured up to 3 cm. Grouped in rich druses, they developed in vein
cavities with galena. Such specimens were very rare and much valued
by collectors. Each crystal was worth a gold ducat. The renowned
Belsazar Hacquet (1739 – 1815), who was born in France, trained as
a doctor and then became a brilliant geologist who centered his attention
on the mineral world of central Europe, complained in one of his treatises
that he actually had to pay two ducats for a single crystal of valentinite
from Příbram. Valentinite is a weathering product of antimony veins,
where it forms as a secondary mineral through oxidation in the upper
parts of the deposits. It forms columnar to needle-shaped crystals. A rich
deposit of valentinite has been found only in one locality — the Constantine province of Algeria. This also happens to be the sole deposit where
it is mined as an ore, with 83 per cent antimony. In all other localities it
occurs in negligible quantities.

Antimony trioxide, $Sb_2O_3$,
orthorhombic. H. 2.5; Sp. gr.
5.7; whitish, yellowish to pale
red, with a diamond or silky
lustre; S. colourless.

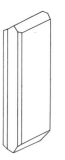

103

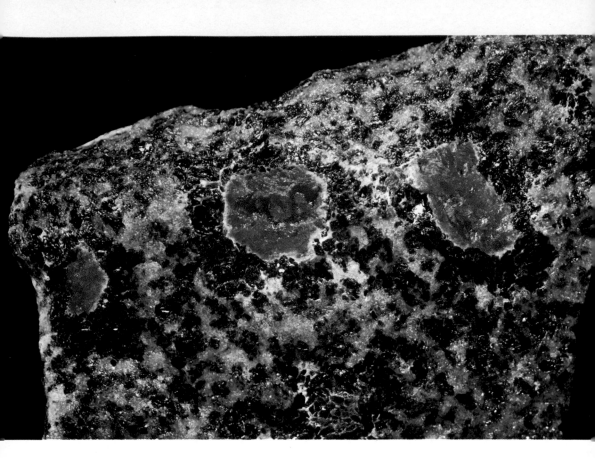

104 Corundum — ruby — Tanzania; actual size of detail 9 × 6 cm.

**Corundum** (104—107)

Aluminium oxide, Al$_2$O$_3$, rhombohedral. H. 9; Sp. gr. 3.9–4.1; transparent (leucosapphire) or variously coloured; traces of chromic oxide cause the red ruby variety, oxides of iron and titanium the blue variety (sapphire); also grey, brownish-red, etc. with diamond lustre, and greasy lustre on less transparent crystals; S. white.

has some valuable transparent, coloured varieties, including the **ruby** and **sapphire**, which have a very long history of recognition as precious stones. The deposits of sapphire in Ratnapura, southeast of Colombo, Sri Lanka, have, for instance, a history of several thousand years. In days of antiquity the robes of important Indian priests were adorned with sapphires. The first written references to ruby deposits come from the 6th century AD. They refer to the mines of Mogok in north Burma. It is probable, however, that rubies were found there much earlier, as in Sri Lanka.

During the Middle Ages sapphires and especially rubies were widely used for the manufacture of jewelry. At that time these stones were ground into the irregular shape of a cabochon — a cut with a round unfaceted top, particularly suitable for semitransparent or opaque stones selected for ornamental purposes. This method was applied because of the limitations of the cutting techniques of that time, and also in order to preserve as much of the precious material as possible. Sapphires and also a ruby cut in this manner can be seen for instance in the Bohemian crown of St. Wenceslas, from 1346. This particular stone is the largest mounted ruby on record. Irregular in shape, it measures 39.5 × 36.5 × 14 mm and weighs approximately 250 carats. The greatest collections of large rubies and sapphires belong to the jewelry of Indian princes and to the Shah of Iran. These precious stones are also usually ground to the irregular cabochon cut.

Corundum forms columnar barrel-shaped crystals, and occasionally crystals which are tabular or needle-shaped. It is, however, mostly granular or compact (**emery**). Corundum deposits occur most frequently in metamorphic rocks and alluvials. For instance the coarse-grained marbles in contact with granites and pegmatites form the parent rock of the Burmese rubies. In Sri Lanka, on the other hand, rubies and also sapphires are mainly found in alluvials. Other deposits of rubies are in Thailand, India, Borneo, Australia, North and South Carolina and Georgia in the USA and in Madagascar. New deposits have been discovered in Kenya, Tanzania, and Cambodia, where the deposits near Pailin look very promising. Attractive, though not particularly transparent, rubies have been found also in the neighbourhood of Prilep in Yugoslavia, where they occur in dolomite marbles and often form rather unusually shaped crystals. The recently discovered Norwegian deposits sometimes yield ruby crystals of high gem quality. But these occur only rarely in nature. The beautiful colour of a ruby originates from the presence of chromic oxide. The richness of colour is particularly distinct in flawless specimens, and these are very valuable.

Asteriated rubies are interesting crystallographically, and highly prized. When viewed from above by reflected light, an asteriated mineral displays

105 Corundum — sapphire — Ratnapura (Sri Lanka); 6.5 × 2 cm.

a six-pointed or a three-pointed star, which is particularly noticeable in stones cut into a lenticular shape.

The coloration of sapphires is caused by the presence of oxides of iron and titanium. The most beautiful examples come from the extensive placers of Sri Lanka, with Ratnapura the central point of commerce. Sapphires also occur in Thailand, Kashmir, Australia and in Montana in the USA. Smaller deposits are found in China, Bohemia and Madagascar. In Tanzania and Cambodia they occur with rubies. Asteriated sapphires are again considered particularly interesting. Asterism is exhibited by sapphire crystals more commonly and to greater perfection than by ruby crystals.

Other coloured gem varieties of corundum are found in nature. White **leucosapphire**, for instance, is used as an imitation diamond, and yellow corundum (or yellow sapphire) and the rare mauve corundum are also highly prized.

Corundum is a highly prized gemstone and also an important mineral for industry, mainly because, with the exception of diamond, it is the hardest mineral known (12). The richly coloured red ruby, of 'pigeon's blood' colour, is the most valued and also one of the most popular of precious stones. Emery is used as abrasive and polishing material. The manufacture of artificial corundum is necessary for industrial and jewelry purposes. Synthetic rubies (13) can be used for instance in lasers.

106 Corundum — cut yellow sapphire (Sri Lanka) and leucosapphire (Sri Lanka); yellow stone 14.4 × 8.7 mm; weight 10.69 carat.

107 Corundum — cut ruby — Mogok (Burma); 16.5 mm in diameter; weight 27.12 carat.

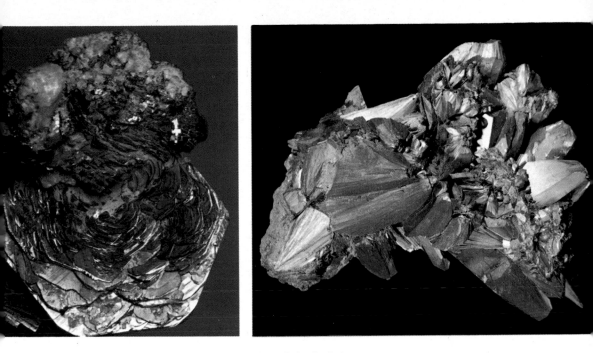

108 Haematite — St Gotthard (Switzerland); 7 cm.

109 Haematite — Bludná (Czechoslovakia); 7 × 6 cm.

**Haematite** (108 — 111) was mined in the ancient world. There is even a reference to it in Virgil's *Aeneid* where he admires the beauty of the rhombohedral, brilliantly coloured crystals of the island of Elba. The colouring occurs through the refraction and dispersion of light in the surface layer of the mineral. Lump haematite was also used for ornamental purposes, and was particularly popular with the Babylonians and Egyptians. The British and Czechoslovakian **kidney ores** (a variety of haematite which forms reniform aggregates) were used for making precious stones for rings. Haematite occurs in the most varied forms and in an extensive range of colours. But the colour of its streak is always red. The most common variety is fine-grained **ochre**. **Oolitic haematite** which originates through deposition on the seafloor, is of particularly high commercial value. **Iron mica** is foliated haematite which occurs in agglomerates commonly in association with quartz. If stratified and rock-forming, it is called **iron mica schist**, or **itabirite** (thus named after the deposits in the Itabira range, Brazil). Haematite develops in nature under the most varied conditions. It is, after all, one of the most abundant of minerals. The red colour of soil, gravel and whole rocks is caused by its presence. The name 'haematite' is derived from the red colour (from the Greek 'haima' — blood). Apart from sedimentary rocks, it also occurs in metamorphic deposits, and in small quantities is found in hydrothermal veins, in magmatic rocks and in oxidation zones.

Iron oxide, $Fe_2O_3$, rhombohedral. H. 6.5; Sp. gr. 5.2–5.3; commonly red, when crystallized dark-grey, often with shady colours, and metallic to submetallic lustre; S. red.

There are many haematite deposits. The richest mines are in Sweden, and the 'Iron Ranges' around Lake Superior in the USA, and the USSR. The so-called 'Alpine roses' form beautiful clusters of flat crystals in Alpine deposits. As an iron ore, it contains approximately 70 per cent metal. The red-ochre variety is used in making pigments and in polishing and grinding.

110 Haematite — Rio Marina
(Elba, Italy); 7 × 7 cm.

111 Haematite — Shabrovskoi
(Urals, USSR); 7 × 7 cm.

**Quartz** (112—139) was well known to the ancient world, especially the purest variety — rock crystal. The name 'crystal' is of Greek origin and means 'ice'. Rock crystal (clear quartz) was thus named because it resembles ice, and remains cold for a long time in warm conditions, unlike glass for instance. The reason for this is that quartz is a very poor conductor of heat. Ancient Romans knew this too, and wealthy patricians often had large crystal balls in their homes, which were there to cool their hands. Pliny the Elder was struck by this particular crystal characteristic; he claimed that crystal originated from ice in such freezing conditions that it could not melt even in the most intensive heats!

Silicon dioxide, $SiO_2$, rhombohedral — pseudohexagonal. H. 7; Sp. gr. 2.65; colour — see text; S. colourless, white to grey.

The Greeks and the Romans were well aware of other quartz varieties, though understandably they had no idea that they were all the same mineral. For example, they were familiar with the violet amethyst (from the Greek 'amethystos' — unintoxicating) (112—116), which was considered a remedy for drunkenness. So it was used in the production of expensive amulets and also as a decorative stone. Another variety with a name of Latin origin is morion, which is nearly black. Pliny used the name for black jasper and onyx from India.

Quartz served man much earlier than this, firstly because it was so widely abundant and resistant to weathering and all external influences. It was without doubt the first mineral to be used by man at the beginning of his existence as a weapon and an essential tool in his fight for existence. Implements made of flint (a variety of quartz) (137) exist from the beginning of the Old Stone Age (Palaeolithic Period) to the end of the Neolithic period, which was the beginning of the 2nd millennium BC.

Quartz has been useful to man through history, and is still so today. The English name 'quartz' comes from the German 'quarz'. Prehistoric man valued flint for its toughness, resistance to wear and for its sharp cutting edges. Gradually through the ages, a number of other excellent qualities of quartz were discovered, so its usefulness grew and today it is widely used in modern industrial manufacture.

This is not all. Even in ancient times miners were amazed at the beauty of quartz crystals. As cultures advanced, so the popularity of quartz and its varieties increased; its beautiful shapes and splendid colours caused it to be very much in demand as a precious and ornamental stone. The magnificently cut goblets from the Middle Ages which were created by famous masters of that era (especially Italians and Germans) command our admiration even today. The raw material for these works of art was supplied mainly by the Alpine deposits.

It is apparent from the jewelry and ornamental objects of the Middle Ages and the Renaissance that the various quartz varieties were extremely popular stones. Their very impressive perfect transparency and strong lustre made them much in demand, and they often replaced other, more rare and more valuable precious stones.

Quartz appears in the mineral kingdom in the most varied forms, either as crystals or massive aggregates. Apart from the clear, glassy **rock crystal** (121—126) there are many varieties of distinctive colour. **Amethyst** is a very popular crystal variety, with a beautiful purple to violet colour. It owes its colouring perhaps to iron admixture and radioactive decay. The colouring of quartz stones from some deposits is not constant even in daylight, and fades fairly quickly.

112 Quartz — amethyst — Capnic (Romania); crystal 3.5 cm.

113 Cut amethyst — Uruguay; 43.2 × 35.0 mm; weight 234.9 carat.

114 Cut amethysts — Brazil; actual size of the largest 51.9 mm in diameter; weight 361.5 carat.

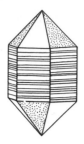

Citrine (117—118) is crystallized quartz, with a pale yellow colour caused by the presence of iron trioxide. **Smoky quartz** (119—120; 129—130) is also a crystallized quartz variety, distinguishable by its smoky-brown to brown-black colour. The black, almost non-transparent variety of smoky quartz is **morion**. The colouring of both these varieties is probably caused through silicon ions being released from their positions in the crystal structure during radioactive decay.

The most popular massive quartz variety is **rose quartz** (127), in various shades of pink — from pale pink to rose-pink, sometimes with a violet hue. This colour is due to the presence of manganese oxide and fades when the mineral is heated to 575 °C. Prolonged exposure to the atmosphere also caused the colour to fade. Until recently, rose quartz was only known in massive form, but a few years ago deposits were found in Brazil where it occurs in crystal form.

Red, brown or yellow, ·**ferruginous quartz** (128; 132—133) is massive, granular or crystallized. **Milky quartz** has no impurities, but is clouded to a milky-white colour and is a veinstone. The milkiness is caused by the presence of a multitude of tiny air cavities. The impure varieties of a wide range of colours are classed as **common quartz. Hornstone** (138) is a mixture of quartz and chalcedony, and so is the similar brownish-black **flint.**

**Prase**, named from the Greek 'prasios' — leek green, owes its colour to the presence of many fine actinolite needles. The pleasant shade of colour made it a desirable stone in ancient times in the manufacture of gems and mosaics, when it was used much more than it is today. Prase was originally called 'the mother of emeralds' and was assumed to be the parent rock of emeralds. **Sapphire quartz** (incorrectly called siderite) is lump, blue-green quartz, whose coloration is caused by the presence of numerous needles of fibrous blue amphibole-crocidolite.

The fibrous, banded quartz varieties are well known in the jewelry trade and are very popular. Their pleasing colouring and lustre, which improve with cutting and polishing, are caused by amphibole fibres, especially of crocidolite. According to their colouring, there are grey-green **cat's-eyes** (136), blue-grey **falcon's-eyes** (136), and **tiger's-eyes** (6; 135—136) with yellow fibres of weathered crocidolite.

**Aventurine quartz** (134) is an exception. It is massive and contains a multitude of minute scales of mica or haematite, which are responsible for the characteristic sparkle of the mineral. It is of interest that imitations of this stone existed before the stone itself was discovered. It so happened that in the workshop of a Venetian glassmaker, Murano, glass material of exceptional lustre caused by the inclusion of minute copper flakes, was created almost by accident ('per avventura' as it was said in Italian). This material was named aventurine and was sold under this name; when later on a mineral of similar appearance was discovered, it was given the same name.

115 Quartz — amethyst —
Banská Štiavnica (Czechoslovakia); 9 × 8 cm.

116 Quartz — amethyst —
Guerrero (Mexico); 6 × 6 cm.

117 Quartz — cut citrines — Brazil; actual size of largest 54.6×49.1 mm; weight 374 carat.

118 Quartz — citrine — Rio Grande do Sul (Brazil); 14×9 cm.

114

119 Smoky quartz —
St Gotthard (Switzerland);
large crystal 11 cm.

120 Cut smoky quartz —
Switzerland; 60 mm in diameter; weight 764.45 carat.

121 Clear quartz — rock crystal — Dauphiné (France); 12.5 × 7.5 cm.

122 Clear quartz crystal — Le Bourg d'Oisans (France); large crystal 12 cm.

123 Quartz — Uri (Switzerland); 5 × 3.5 cm.

124 Quartz — Marmarosh diamonds — Marmarosh (Ukraine, USSR); crystals 4—8 mm.

125 Quartz — Minas Gerais (Brazil); 8.5×5 cm.

126 Quartz — Hot Springs (Arkansas, USA): 11×8 cm.

Apart from the crystallized, granular and massive quartz varieties already mentioned, there are many others, which are distinguishable by the varied growth of their crystals. Individual crystals and crystal aggregates, which originated mainly from hot solutions, and which differ in their form in various ways, belong to this group. **Cap quartz**, for instance, is crystallized quartz, in which small layers of intruding matter, such as mica, formed during its interrupted growth. Such a crystal can be split with one blow into many little 'caps'. **Star quartz** consists of

127 Rose quartz — Brazil; 12 × 9 cm.

individual crystals grouped in a star-shaped cluster. The so-called **token quartz** (128), which occurs mostly as a veinstone in Slovakia or in fissures of crystalline schists in the Swiss Alps, has crystals which form in the following manner: the original columnar crystal narrows at the end and another, somewhat wider crystal grows into it where it terminates; it is usually bounded on both sides and is always oriented the same way as the slender, lower crystal.

128 Token quartz — Banská Štiavnica (Czechoslovakia); large crystal 7 cm.

129 Smoky quartz — Dolní Bory (Czechoslovakia); 9,5 × 6 cm.

130 Smoky quartz — detail; actual size of pictured section 4 × 3 cm.

**Porphyritic quartz** occurs as crystals formed from the smelt of some porphyritic rocks which are bounded only by rhombohedral faces but not prismatic faces (called dihexaedras). **Babylonian quartz** develops in shapes which are narrowed at the top, **stone-wall quartz** forms crystals which are intergrown on parallel lines and which display on their transverse side patterns resembling stone walls. **Honeycombed quartz** occurs as a covering pseudomorphism after the dissolution of original minerals, which left behind only surface crusts of quartz.

Quartz is widely distributed in nature as a filler of individual veins, which often contain gold or other ores. It is commonly grey-white, at times with a prominent greasy lustre, and is microscopically granular with interlocking lobed minute grains. Usually it fills numerous cracks of rocks, particularly in the vicinity of magmatic intrusive rocks.

Some quartz varieties, often in association with chalcedony (see jasper), formed in the cavities of extrusive igneous rocks, particularly melaphyres. Components of the igneous rocks, were decomposed after their consolidation by hot water and a gel of silicic acid, thus freed, filled the hollows.

131 Quartz — Cínovec (Czechoslovakia);
16 × 10 cm.

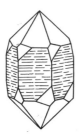

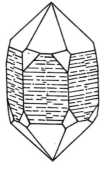

Quartz is not only a substantial component of numerous igneous, metamorphic or sedimentary rocks, but often is almost their sole constituent — as in the case of the most common of these — sand and sandstone. Quartzite, for example, contains 95 per cent fine-grained quartz. It fills either huge cracks in the earth's crust (as in the Bavarian quartz wall, which runs along the Bavarian Forest and Bohemian Forest, or occurs as a stratified, fine-grained rock which transforms into crystalline schists, or, finally, as a fine-grained sandstone with siliceous cement. Quartzite is a sought-after material for the manufacture of heat-resistant material essential to the lining of blast surfaces.

**Lydite** is a similar rock, usually black, grey-black or black-grey through the presence of a large proportion of carbon. It is a fine-grained to compact quartz rock with a splintery fracture, formed uncommonly by sometimes still visible needles of fungi or the remains of radiolarians (microscopic marine animals). There are known occurrences in Bohemia in the range of rocks from the Proterozoic Era between Klatovy and Brandýs nad Labem. The rich black variety of lydite (lapis lydius), found in Lydia, Asia Minor, was familiar to the ancient Greeks and Romans. It is a known 'touchstone' or 'assaying stone', used for testing the purity and the constituents of gold and its alloys.

Quartz crystals are normally very pretty. They are mostly hexagonal prisms, terminated by two rhombohedra, which are sometimes symmetrical. The angle between the prism faces always remains at 60°, regardless of the crystal's development. The prism faces are usually horizontally striated. The striation is due to ridges alternating between the prism and the rhombohedron during the growth of the crystal. The conditions during crystallization are by no means always ideal. Such side effects as pressures or increased flow of solutions can cause crystals to grow distorted. Yet the angle at which individual faces intersect always remains constant. Nicolaus Stensen, a Danish doctor and geologist, studied the angles of quartz and published the result of his investigations in 1669, formulating them as the law of constancy of angle on crystals of one mineral. With this he became the founder of scientific crystallography.

Other types of quartz crystals are rarer than crystals with prismatic and rhombohedral faces. Yet one often encounters a crystal where both rhombohedra are completely symmetrical, which results in a shape similar to a hexagonal double-pyramid and gives the illusion that the quartz symmetry is transferring to the hexagonal. Such crystals occur in some porphyries, where quartz has crystallized at a temperature higher than 573 °C and, after a fall in temperature, still retained the external bounding of the so-called upper quartz. This is why quartz is considered a pseudohexagonal mineral. Such crystals are sometimes shorter, so there is a dominance of rhombohedral faces while the prismatic ones almost disappear. Quartz has no cleavage and has an uneven or conchoidal fracture. It is also rather brittle.

There are some giant specimens among quartz crystals. These come particularly from Madagascar, where individual crystals have been found with a circumference of several metres.

One can frequently find in quartz quite a number of small cavities, partially or fully filled with the parent solution from which quartz crystallized. Study of such cavities tells much about the conditions under which quartz developed in individual deposits.

Quartz is the commonest and the most widely occuring mineral of the earth's crust, responsible for 12 per cent of its composition. Apart from this there is about 47 per cent silicon dioxide in various forms of silicates, which are the chief rock-forming mineral. Quartz is a substantial component of many igneous rocks (for instance the greyish grains of granite) and of metamorphic rocks (gneisses, mica schists and quartzites).

Fairly large quartz crystals are found mainly in cavities of coarse-grained granites, in pegmatites, in some ore-veins, and in crystalline schists. In ore-veins they commonly develop in gangue, though entire quartz veins are also common. Crystals of clear quartz usually occur in crevices of schists, but also in igneous rocks, such as melaphyres, often jointly with agates. Amethyst too can be found in cavities of igneous rocks, where it forms geodes — round-shaped stones with a hollow centre. It is also recovered from the gangues of ore-veins and from quartz veins. Smoky quartz and morion occur similarly. Rose quartz usually forms in pegmatites.

132 Ferruginous quartz — Westphalia (Federal Republic of Germany); 10 × 8 cm.

Hornstone occurs in ore-veins as a mineral of inorganic origin; flint occurs as nodules of organic origin in Mesozoic sedimentary rocks. As a mineral highly resistant to weathering and decomposition it often finds its way into sand and gravel beds or alluvials. Quartz is therefore in great abundance in many sedimentary rocks. The attractive pebbles and stones of every conceivable colour found on the shores of rivers, lakes and seas, and formed chiefly by quartz, are a reminder of this fact.

It is obvious that quartz originates in nature in many varied ways. It crystallizes anywhere between the highest temperatures of volcanic magma and the normal temperature of the earth's surface, from hot solutions or solutions comparatively cooler.

The most magnificent quartz crystals, especially of clear quartz, come from Alpine schists. In 1719 an enormous cavity was discovered in Zinggenstock, near Grimsel in the Bernese Alps of Switzerland. It has been called the 'crystal cellar', for more than 100 tonnes of the most beautiful clear crystals have been extracted from it. Some pieces weighed

133 Ferruginous quartz — Hořovice (Czechoslovakia); 7 × 6.5 cm.

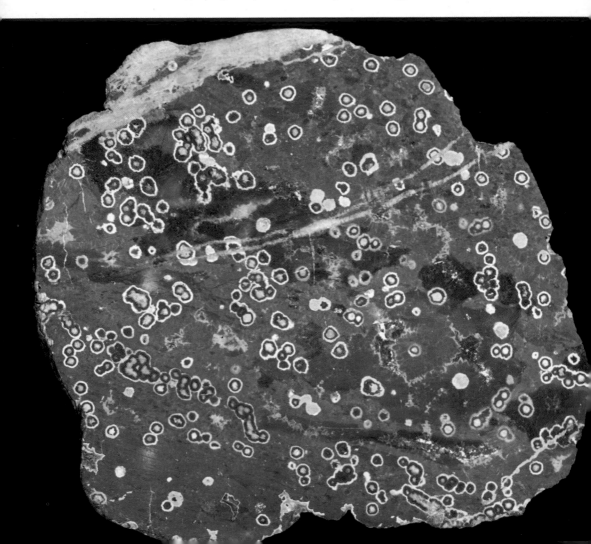

as much as 800 kg. The richest crystal deposits at the present time are in the eastern and northeastern part of Madagascar, southern Brazil (the surroundings of Cristalina in the mountain range of Serra dos Cristaes in the state of Goiás), and also Arkansas in the United States. In Arkansas in the Hot Springs area a cavity measuring $2 \times 10$ metres was discovered, from which wagon loads of clear quartz were recovered. In Madagascar rocks up to 50 kg in weight are extracted; the largest known crystals of rock crystal come from these deposits. The Brazilian crystals are sometimes more than half a metre in thickness. In the Carpathian Mountains of the Soviet Ukraine there are deposits of little crystals, called 'Marmarosh diamonds' from Tertiary sandstones and schists.

Collectors still show tremendous interest in druses of rock crystal, for they are indeed the most wondrous creations of nature. Rock crystals, grown to a base, are most commonly columnar, and often decorated with natural etchings. Usually they are completely colourless, or have brownish

134 Quartz — aventurine — Styria (Austria); $8 \times 5.5$ cm.

shades and are therefore transitional to smoky quartz. Often they contain interesting inclusions, such as the scales of green chlorite, needles of actinolite or rutile, or other minerals. Cracks in many of the crystals display the play of rainbow colours, which results from the interference of light.

Rock crystal, of course, occurs in nature less frequently than common quartz. It is, however, fairly widespread in sedimentary rocks, such as limestones, and in cavities of pegmatites, ore-veins and in the previously mentioned cracks of crystalline schists (called veins of the Alpine type). Rock crystal generally occurs here together with adularia, chlorite and titanic minerals. All these are very much in demand and are sold to tourists at high prices, particularly the stones from the oldest and best known rock-crystal deposits in the Alps. Through weathering and alteration in the parent rocks, rock crystals reach fluvial deposits, where they lose their perfect crystal shape and turn into cobbles. Only when such a cobble is broken, does the transparency typical to the rock crystal prove its identity. Many of the finds of rock-crystal pebbles in various fluvial deposits and gravels of rivers, which run down the north Alpine slopes, are of such an origin. Such cobbles of rock crystal are even present in the Rhine in Holland. Up until the 1860s, the Alps and their immediate surroundings were the only known area to produce large crystals and pebbles of rock crystal.

The crystals of rock crystal found in cavities of the famous snow-white Carrara marble in northern Italy are considered to be the most perfectly bounded specimens; they very much resemble the previously mentioned Marmarosh diamonds, but exceed them in size and beauty. Herkimer diamonds are similar; they are loose crystals of rock crystal from Herkimer, near New York. Their distinctive transparency and strong lustre are a true reminder of real diamonds.

The most exquisite examples of amethyst druses with perfectly developed and coloured crystals of gem quality are found in extrusive rocks in Brazil. Whenever they are found, it is always a great occasion, for they are indeed exceptionally beautiful. They occur either directly in melaphyre cavities, or develop as individual, loose crystals or as rock fragments in fluvial deposits and in topsoil, which they reach after the weathering of the original parent rock. The richest finds have been made in the southernmost Brazilian state Rio Grande do Sul and in adjoining Uruguay, chiefly in the Serra do Mar range. Here they are found in the shape of large round balls, which are often hollow inside (called geodes), with the walls padded with numerous amethyst crystals which often are of an exceptional size. This wealthy deposit was discovered in 1823 by German grinders, who had emigrated to South America from Idar-Oberstein. Locating and mining amethysts is extremely simple, but demands experience and capability. Long, pointed iron rods are used to prod and pierce the weathered soil of igneous rocks, hillside detritus and fluvial deposits in the search for the round nodules in whose core hide the deep violet amethyst crystals. Some of the discoveries were truly amazing. Unfortunately such crystal caves are rarely preserved as a whole, but are cut and divided into a number of sections and sold to museums throughout the world. Amethysts also occur in other parts of Brazil. In certain areas of Bahia state there are large quartz-veins in siliceous rocks; sizeable, magnificently coloured amethyst crystals, highly suitable for use as gemstones, frequently develop in their cavities.

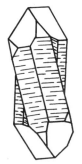

The largest cut and polished quartz is a rock-crystal nodule in the collection of the Smithsonian Institution in Washington. It was cut from a large block of quartz found in Burma, weighing over 450 kg. The nodule is 32 cm in diameter and weighs almost 48.5 kg. The largest quartz worked into a gemstone is also a part of this collection. It is a Brazilian

rock crystal, weighing 7,000 carats. Next comes a pale smoky quartz from California with the weight of 4,500 carats, then dark smoky quartz from Brazil which is 1,695 carats in weight. All these stones are also in the collections of the Washington museum.

The largest amethyst geode was found in 1900 in Rio Grande do Sul in Brazil. It measured $10 \times 5 \times 3$ metres and weighed 7 tonnes. Parts of this geode can be seen in a number of museums. Similar, but far smaller pockets of crystallized or massive quartz varieties occur near Idar-Oberstein in Germany and also in the foothills of the Bohemian Krkonoše Mountains. The wealthiest amethyst deposits, apart from Rio Grande do Sul are in adjoining Uruguay, in Mexico and Madagascar (Betafo), where particularly richly coloured stones are extracted. The loveliest European amethysts come from the Urals (Murzinka) and from the ore-veins of the Carpathian Mountains, especially near Capnic in Romania and near Banská Štiavnica in Slovakia. However, the colour of the amethysts from Banská Štiavnica fades fairly quickly when they are exposed to daylight.

There are valuable deposits of smoky quartz and morion in the Alps, especially in the vicinity of St Gotthard, where they form in pockets in granite rocks. In 1868 a cavity, measuring $6 \times 4 \times 2$ metres and containing 30 tonnes of smoky quartz was discovered in the Swiss canton of Uri. The largest of these specimens are on show in the Natural Science Museum in Bern (the biggest crystal is 70 cm long and weighs 130 kg). Magnificent druses of smoky quartz occur in the Urals (Murzinka), in southern Brazil, Madagascar and in Colorado. Attractive smoky quartz crystals also occur in Czechoslovakia, especially in pegmatites at Dolní Bory near Velké Meziříčí.

Natural citrine is comparatively rare. The most beautiful pieces come from Brazil and the Urals. Most of the citrines available on the market are artificially heated amethysts or smoky quartz. They are sold under the names of 'gold topaz', 'Spanish topaz', 'Madeira topaz', or 'topaz from the Urals'.

The localities of rose-quartz deposits are Madagascar (massive) and Brazil (crystallized). In Europe they form large masses in pegmatites in Bavaria and also in the Urals.

Massive ferruginous quartz occurs in some iron-ore deposits near Hořovice in Bohemia and is found in crystal form at Iserlohn in Nordrhein-Westfalen, West Germany. Green aventurine quartz was in great demand in China and was named 'the imperial stone YU'. Aventurine is also found abundantly in the Urals (vicinity of Miass), in Siberia, Tibet and India, and in Europe in Styria.

Quartz which has been permeated with fibrous minerals is not frequently found in deposits. Cat's-eye and falcon's-eye occur chiefly in Sri Lanka; so does tiger's-eye — very popular and once also valuable. They also come from the Asbestos Mountains of South Africa. Tiger's-eye is also found in China. These stones were widely used for making decorative objects many of which were brought to Europe. In recent years similar deposits have been discovered in Australia. Star quartz was chiefly a Bohemian speciality. Found in Perimov in northern Bohemia, it formed magnificent radially arranged crystals. This deposit is now exhausted and the star quartz can be viewed only in collections. From the Bohemian deposits in Horní Slavkov come also the best examples of large crystals of cap quartz.

There are substantial quantities of flint on the shore of the Baltic Sea, particularly on the island of Rügen. Flint was carried to central Europe mainly by northern ice-sheets, which covered the whole northern region of Europe during the Ice Age. As a constituent of the moraine of the northern glacier, nodules and fragments of flint now occur in gravels in Germany and Poland and are found also in the northernmost regions of

Bohemia and Moravia. Its concretions are of a variable shape, commonly coloured black to black-grey by the presence of carbon. In some localities there are occurrences of flint coloured yellowish or reddish by impurities. Its distinct chalk-white crust is formed by the powdery quartz with an admixture of chalk. The concretions break with a prominent conchoidal or splintery fracture, and display a dull or a faint vitreous lustre. They developed either through the concentration of dispersed silicic acid in limestones into concretions, or through the acid being freed from the needles of siliceous sea fungi, and the quartz shells of radiolarians. Occurrences of flint are also very common on the shores of the English Channel and the Baltic Sea.

Clear quartz (rock crystal), which is a relatively inexpensive precious stone, is today used in making larger art objets. As it is hard and tough, it is a most suitable material for this purpose, and so are the other quartz varieties. Amethyst is most sought after and is widely used for making jewelry and decorative objects. It is prized as a gem for the richness and consistency of the colour of its crystals.

When heated to approximately 250 °C, some amethysts become a rich honey yellow. It did not take long for jewelry manufacturers to take

135 Quartz — tiger's-eye — Griquatown (Republic of South Africa); 9 cm.

advantage of this and to turn the stones into artificial citrines, which are choice precious stones often taken for topaz.

Amethysts, citrines, smoky quartz, clear quartz and sometimes even rose quartz are ground to a brilliant cut, either of elongated, oval or a round shape. Smoky quartz, morion and amethyst are particularly suitable as larger ring stones. Various varieties of quartz are a suitable material for glyptics, i.e. for carving and engraving either with a raised design (cameo), or an incised design (intaglio).

Transparent quartz with interesting inclusions is very popular in the making of jewelry. Such inclusions can be flakes of green chlorite, needles of actinolite, tourmaline or brown rutile — sagenite (flèches d'amour). They are often sold at a comparatively high price to tourists in regions where they occur (such as the Alps). The same applies to nice crystallized quartz specimens, especially specimens of clear quartz (rock crystal), which are often accompanied by other interesting minerals such as adularia, chlorites, or various titanium minerals. Alpine crystal hunters scale the most dangerous places in order to follow the course of quartz veins. They determine the position of vein cavities with a tap of the hammer and proceed to open them with a pickaxe, using all their expertise. These cavities often occur in the most dangerous and inaccessible positions. The crystal hunters are sometimes lucky enough

136 Quartz — cat's-eye, cut and polished (Sri Lanka); tiger's-eye (Griquatown, Republic of South Africa); falcon's-eye (Griquatown, Republic of South Africa); largest stone 35.1 × 24.5 mm.

129

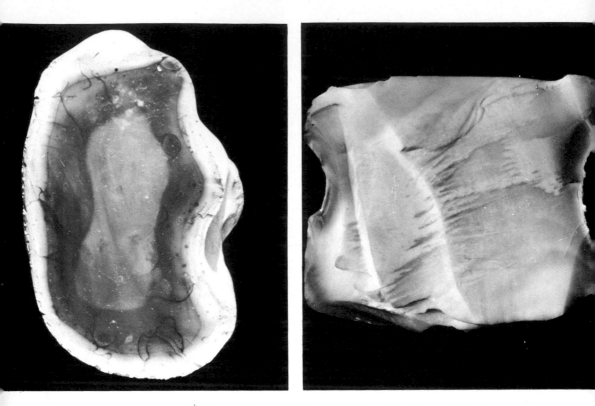

137 Quartz — flint — Heligoland (Federal Republic of Germany); 12 × 8 cm.

138 Quartz — hornstone — Madeira; 8 × 7 cm.

to be the first to discover the breathtaking magnificence of the true crystal caves, whose walls are lined with pure, clear crystals, or with smoky quartz crystals of fairy-tale loveliness. Most of these crystals occur in Switzerland.

Today quartz is an important raw material to the glass and optical industries. The transparent, pure clear quartz is used in the manufacture of optical glass. An important and unusual use of quartz is in radio-engineering. Quartz crystals manufactured artificially are mostly used for this purpose today. They are used for various optical instruments (for polarizing microscopes to identify minerals; in spectroscopy; for astronomical telescopes, etc.). The attention of the world has lately been centred on the synthetic production of quartz, for the time when natural deposits of rock crystal needed for technical purposes will be practically exhausted, is not that far away. Artificial clear crystal is valuable for all the needs of industry, especially for electronic equipment, which demands not only a perfectly pure material, but first and foremost a material with an internal structure which is monoclinic. This demand is not usually met in natural crystals, even those which are perfectly suitable for use as gemstones. Most of them occur as twinned crystals. It is understandable then that the demand for synthetic quartz crystals is exceedingly high in all technically developed countries. In the United States, for instance, the consumption of quartz crystals is about 3,000 tonnes a year.

139 Quartz — petrified wood — Schneeberg (German Democratic Republic); 15 × 11 cm.

**Jasper** (140–144) is one of the precious and decorative stones used in the ancient world. In fact in the past jasper's popularity was far greater than now. Jasper was used for making ornaments and amulets. The ancient Egyptians, Greeks and Romans used jasper. They employed various methods for boring holes into the stones and for engraving different symbols, signs and later even portraits on their surfaces. This was the beginning of glyptic art. Pliny also wrote about the colourful splendour of jasper.

Silicon dioxide, SiO₂, mixture of rhombohedral quartz, cryptocrystalline chalcedony and colouring impurities. H. approx. 7; Sp. gr. variable; red, ochre yellow, green, dark, with greasy to vitreous lustre; S. white to grey.

Jasper's popularity lasted through the Middle Ages. The chapels of the Bohemian Karlštejn castle and the St Wenceslas chapel of Hradčany castle in Prague have walls partially decorated with blood-red jaspers. The decoration dates from the reign of the Emperor Charles IV who, in 1347, had all his favourite places decorated with jaspers. At that time, jaspers and amethysts were being extracted from a vein on the Saxony

140 Jasper — Kozákov (Czechoslovakia); 15 × 9 cm.

141 Jasper — Mokkatam
(Egypt); 9.5 × 5.5 cm.

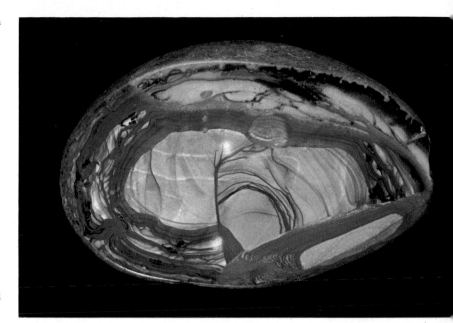

142 Plate of jasper, cut and
polished — Egypt;
9.5 × 6.3 cm.

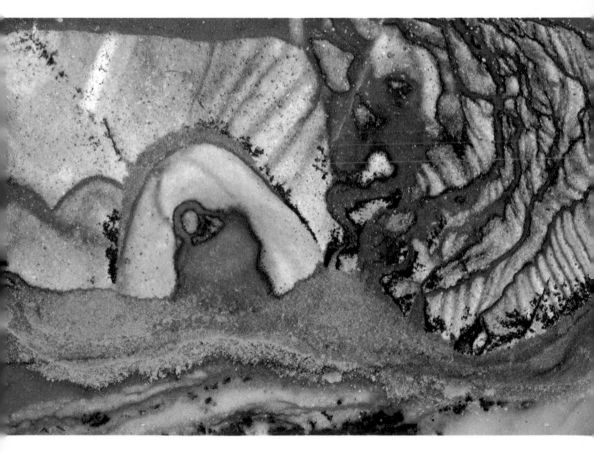

143 Jasper — Bulanda (Iceland); 10 × 9 cm.

side of the Ore Mountains, then transported to Prague for cutting and polishing. Specially selected discs were embedded in mortar and rimmed with gold.

The work of de Boot, the personal surgeon to the Emperor Rudolf II, entitled *Gemmarum et lapidum historia* says a great deal about the popularity of jasper during the Renaissance. De Boot describes, for instance, a large table top from the Emperor's collection, made up of a multitude of coloured jaspers obtained from deposits below the Krkonoše Mountains, which depicted the countryside and was then considered one of the seven wonders of the world. During that period there were many works of art created from the Bohemian jaspers, though not so famous as the one just mentioned.

Hermitage Museum in Leningrad (once the Winter Palace of the Tsars) holds a wealth of objects made from jasper in its halls and chambers. There are pillars and vases measuring several metres, often carved from a single piece of the mineral, made by craftsmen-cutters in the well-known Russian workshops of Petrodvorets and Ekaterinburg (now Sverdlovsk). Rich deposits near Orsk in the Urals yielded the minerals for this purpose. Jaspers found in that locality are still known for their striking gaiety of colours. Magnificently carved Indian jaspers can also be found in the collections of Indian maharajahs.

Pliny used the term 'jasper' for the lighter types of this mineral. At the same time he named a number of the coloured varieties; the dark to black jaspers from India were originally classed as 'morion'. This name was later applied to dark, smoky quartz. Green jasper with regular, blood-red haematite stains was then called heliotrope (from the Greek 'helios' — sun, and 'trepa' — to change), for in the ancient times this stone was used for observing the movements of the sun.

Jaspers occur frequently in nature; they are inseparable companions of agates and chalcedony. They develop mainly through crystallization of hot solutions in cracks of igneous rocks, usually melaphyres. The mineral components of these igneous rocks were, after consolidation of the rock, decomposed again through the effect of hot solutions. The siliceous gel, which was freed by this, was carried into the cracks. There it crystallized into fine-grained to compact opaque quartz together with chalcedony; these are the basic components of jaspers, except for additional natural mineral colouring substances. Powdery haematite dyes them red, limonite paints them brown to yellow and the mossy chlorite aggregates turn them green. Sometimes jaspers are formed by the metamorphism of sedimentary rocks caused by the hot melaphyre lava. Through the weathering of the parent rock, jasper is frequently transported to hillside detritus, which is a good place to search for this mineral.

Jaspers also form variously shaped concretions in some sedimentary rocks, for instance in Egyptian limestone, and are freed by their weathering. They occur generally in the form of a flat tuber. These interesting, chestnut-brown or coffee-brown jasper cobbles with a strong lustre, are also called 'Nile quartz' and they have a somewhat unusual surface produced by grains of sand being hurled against them by the desert winds. When cut and polished, they are indeed an attractive sight, for they are usually beautifully banded and decorated with black branching tree-like shapes (dendrite shapes), which resemble fossilized moss. They are used in making souvenirs, ornamental objects and articles for the tourist trade.

Jaspers are very colourful stones. A one-colour specimen is more or less a rarity. Their very variable structure, which may be banded, layered or mossy, is emphasized even more by their gay coloration. The distribution of the colours is hardly ever regular, and they usually come in cloud-shaped, or ribboned patches, with one colour gradually merging into another. Agates and chalcedony often need to have an addition of artificial colouring, but jaspers have been created by nature with a permanent natural colouring, thus ensuring their suitability for use as a raw material without the necessity for further artificial addition of colour. The finest examples of jasper with a prominent colour variety are found in the same deposits as chalcedony and agates. In many of these localities they are, in fact, more abundant.

The Krkonoše Mountains in Bohemia yield abundant magnificent jaspers (at Kozákov for example). Notable occurrences are in the southern Urals, from where jasper boulders weighing many tonnes have been recovered. German jasper is also well known, especially from the Rhineland (Idar-Oberstein) and many localities in Saxony (the valley of the river Müglitz and surroundings of the town of Karl-Marx-Stadt). Large deposits also occur in India, Brazil and in the Egyptian deserts near the Nile. Heliotrope is found in many of the Indian deposits. It occurs together with other jasper varieties more or less in the same places as agates and chalcedony; together with them it is treated by the local gem-cutters, or exported to Europe and China. Fine examples also come from the Urals, from the Tirol and from the foothills of the Krkonoše Mountains in Bohemia, where they are considered as popular gemstones. New deposits of jasper have also been found in Madagascar.

Jasper as a raw material is suitable for the manufacture of a large variety of objects, especially ornamental ones, because of its chemical and physical resistance; because it is easily accessible; because it occurs in large, unblemished pieces and last, but not least, because of its attractive appearance. This material has been used for making a variety of small objects since ancient times, especially elegantly curved handles, clips and slides, ornamental buttons, etc., and also more valuable objects, with exquisite carvings and engravings, such as brooches, rings, earrings, necklaces, pendants, etc. The Bohemian foothills of the Krkonoše Mountains were a world-renowned centre for cutting and working these stones. This industry thrived until the arrival of jasper, agates and quartz from deposits in Brazil and Uruguay. They came to the workshops of Idar-Oberstein in the Rhineland; superior in size, and more abundant in numbers, they soon pushed the Bohemian products out of the European market. Jaspers are now widely used in the USSR and also in parts of North America, mostly for the manufacture of dishes, goblets, paperweights and mosaics. The finest pieces of jasper are also used for making gemstones. Egyptian jaspers serve for making souvenirs and ornamental objects, mainly for the tourist trade.

144 Jasper — Camamba (Madagascar); 10 cm.

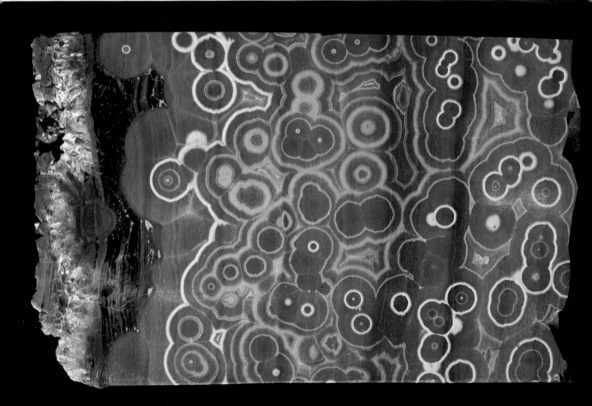

**Chalcedony** (145—157) was a very familiar mineral to ancient civilizations. In those days it was a most popular and sought-after stone especially in Egypt, Greece and Asia Minor. The ancient Egyptians excelled in the art of gem cutting, engraving and polishing. Specimens of chalcedony scarabs, inscribed with hieroglyphs, representing the beetle *Ateuchus sacer*, which were worn as amulets, have been preserved to the present day. They have been found mainly in tombs of mummies together with other jewels and precious stones. Talented artists of the ancient times engraved chalcedony and its varieties, thus gaining a permanent place for this stone in the history of art.

Silicon dioxide, $SiO_2$, rhombohedral — cryptocrystalline. H. 6.5; Sp. gr. 2.59–2.61; variously coloured, most commonly milky-grey, yellow-red (carnelian, sard), or green (plasma), often speckled, with greasy to vitreous lustre; S. white to grey.

Chalcedony was named after the ancient city of Chalcedon (now Kalkidoy), near Istanbul, where it was discovered. There are many varieties of chalcedony, including **carnelian** (152) and **plasma.**

During the Middle Ages **chrysoprase** (153), a pleasantly apple-green chalcedony, whose name is derived from the Greek 'chrysos' — golden, was quite a valuable ornamental stone. During the reign of the Emperor Rudolf II, the demand for chalcedony was so great, that the Emperor sent personal collectors to the deposits beneath the Krkonoše Mountains, including collectors from foreign lands, especially Italians, who were specially commissioned by him and given many privileges.

Chalcedony is a very interesting variety of quartz, which differs from other varieties by a number of its characteristics and the mode of its origin, for it comes from gelatinous siliceous matter. Externally chalcedony appears to be a compact mineral, but an examination under the microscope shows that it is in fact an aggregate of extremely fine, closely packed fibres, which form either layers or round aggregates. As a microcrystalline quartz, chalcedony has a number of varieties, which differ in their optical properties.

Chalcedony usually forms layers with a reniform surface and an even to splintery fracture. It very often forms with opal and a variety of colouring minerals, such as haematite, limonite and chlorite. The fine pores in between the individual fibres, when they are large enough, enable chalcedony and its varieties to be artificially coloured. Inorganic dyes are now used more commonly than aniline dyes for adding colour to chalcedony. For example, chalcedony turns blue in a ferruginous solution and through the effects of yellow prussiate of potash; it turns red in green vitriol solution and immersion into ammonia and with annealing, which gives rise to ferric oxide (a dye similar to natural haematite); it turns yellow through the effects of hydrochloric acid, etc. The stones are usually immersed into the dye solution for a period of several weeks, and are heated to 50—60 °C. In this manner a wide range of colours and hues desirable for jewelry purposes is reached. Chalcedony found in Brazil and Uruguay is the most suitable for artificial dyeing. The colour of some of the other chalcedony stones, such as the Bohemian ones, is always natural, because of the density of their fibrous structure. As the old stoneworkers remark, they are too hard, which means insufficiently porous, and this causes them to reject dye solutions. But their natural colours, on the other hand, are generally constant, which is a great asset.

The methods of artificial dyeing were once the carefully guarded secrets of the Italian engravers. The secrets were revealed by accident only at the onset of the 19th century to the engravers in Idar-Oberstein on the Rhine, where the gem industry, fed by the rich deposits of numerous chalcedony varieties, agates and quartz located in that area, had been growing since the 16th century. The acquisition of the formulas of the artificial dyeing process gave a great boost to the Idar industry, which then perfected its colouring methods and started to produce the most varied shades of the stones. In the numerous grinding plants, which used cheap water power, tasteful and relatively inexpensive ornamental

145 Chalcedony — Tri Vody (Czechoslovakia); actual size of illustrated section 15 cm.

objects were made by hand from the minerals and sold to all parts of the world. This is how the knowledge of the artificial dyeing process of chalcedony brought prosperity to the grinding plants of Idar-Oberstein, which today are world-famous.

The grinders found that the chalcedony raw material could be used for a wide variety of objects. First they concentrated on small articles, such as elegantly carved handles, slides, ornamental buttons, grips, etc.; later objects of greater value were produced, often exquisitly dyed and engraved. Articles such as brooches, signet rings, ear-rings, necklaces and pendants began to appear on the production line. The increase of business naturally made greater demands on the supply of the raw material and they began to import minerals from South America. The varieties of chalcedony are divided according to their colouring and structure. A number of these contain admixtures of quartz and opal, which often grow into one another and replace each other. Some chalcedony varieties

146 Chalcedony — Ruskov (Czechoslovakia); 10 × 7.5 cm.

are greatly translucent; others, particularly the ones deeply coloured, are sometimes completely opaque. There are a great number of colour varieties.

Collectors' names for the different coloured varieties are often quite unnecessary from the mineralogical viewpoint, but they are commonly known and commonly used. **Carnelian** (10) is tinted red with oxides of iron. Its own variety, the orange **sard**, is valued as a precious stone. **Plasma** (151) is coloured green by minerals containing chlorite. **Moss agate** (154—156) is a chalcedony containing small dendrites, which are either oxides of iron or manganese (brown or black), or chlorite (green). **Cacholong** is a chalk-white mixture of chalcedony and opal. Its name of Mongolian origin means 'a beautiful stone'. **Chrysoprase** is an apple-green chalcedony, coloured with nickel oxide. The name is derived from the Greek 'chrysos' — golden. The common chalcedony forms grey, bluish, or yellowish transparent aggregates.

147 Chalcedony — Sirk (Czechoslovakia); 10 × 7 cm.

148 Chalcedony — La Speranza (Sardinia, Italy); 16 × 13 cm.

The richest and largest occurrences of chalcedony are in cracks and holes of melaphyre and amygdaloidal lava, which solidified on the earth's surface as lava-flows. The numerous holes in these rocks were formed gradually by escaping gases, and were filled by the secondary siliceous gel. Today such chalcedony concretions are found not only in the original rock, but in gravels and in topsoil, which developed through the weathering of the igneous rocks. They also occur in the alluvials of streams and rivers, which often carry them far away.

Primary and secondary deposits of this type occur in Brazil, Uruguay and in India, with particularly rich deposits in the Deccan. The main European localities are Idar-Oberstein in the Rhineland in Germany, the foothills of the Bohemian Krkonoše Mountains, Iceland and the Faeroes. Uruguay and the adjoining Brazilian state of Rio Grande do Sul are today the principal producers of chalcedony; lumps weighing several tonnes have been mined there.

Chalcedony can also originate in most deposits from the decomposition of silicates. This frequently happens in the surface layers of ore deposits, as in Styria (Erzberg), in Slovakia and on Sardinia (La Speranza). Beautifully shaped small chalcedony stalactites and stalagmites can be seen in these localities, with their transparently white, or pale blue colouring, and often forming miniature caves. Plasma results similarly through the weathering of serpentines, frequently in association with

149 Chalcedony — Tampa Bay (Florida, USA); 10 × 8 cm.

150 Chalcedony — Chihuahua (Northern Mexico); 15 × 8 cm.

opal. Chalcedony thus originates as a secondary mineral. Chalcedony can be found in various other places, such as in clayey siderites, in coal deposits, or in siliceous concretions of sedimentary rocks (some of the cacholongs).

Carnelian occurs mainly in Brazil and India (the loveliest carnelians come from the hinterland of Bombay). In Europe it is found in the foothills of the Krkonoše Mountains of Bohemia and in Transylvania, Romania. Carnelian was discovered in Arabia and Egypt in ancient times, and so was sardonyx in Asia Minor. The stones found in those deposits were at that time highly valued as precious gems. They are still being exported to be worked in Europe from these areas, and also from Brazil and India. The most noted occurrences of moss agate are in India (green), the Urals (green and black), and the town of Mocha (or Al Mukna) in the Yemen (black). The latter were exported to Europe under the name of 'Mocha stones'.

Cacholong occurs abundantly in the Mongolian deserts, in the basalt rocks beneath the Krkonoše Mountains, and in cavities of quartz concretions of the Mesozoic Era in the Moravian Karst. The best deposits of chrysoprase are in Koźmice, Polish Silesia.

Some chalcedony from Uruguay is rather interesting, for it contains liquid remains (mother lye). Such specimens are called **enhydros** (157), which is derived from the Greek 'udor' — water.

Many important new deposits of chalcedony and its varieties have been recently discovered in America. The most noted ones are deposits in the basalts of the Nova Scotia peninsula, Canada, centred chiefly

151 Chalcedony — plasma — Hrubšice (Czechoslovakia); actual size of detail 6 × 6 cm.

along the northwest shore. The deposits in the basalts of the northern shores of the Great Lakes, Ontario, are of a similar type. Chalcedony in the form of petrified (silicified) wood in substantial quantities is found in the Red Deer Valley, Alberta, Canada, also in Eden Valley (Sweetwater County) in Wyoming (USA), and also in the well-known natural park, the silicified forest near Holbrook, Arizona, USA. Chalcedony occurs commonly in association with the crystal varieties of quartz in many localities in the southern regions of British Columbia (Canada). The most magnificent pseudomorphs of chalcedony after corals and seashells come from the seabed of Tampa Bay, Florida (USA) and to the north. Coral colonies which have altered into chalcedony found there

152 Chalcedony — carnelian — Almas (Romania); 9 × 9 cm.

often reach up to 60 cm in diameter, yet the original external appearance is preserved. Only the hollow interior of such colonies is usually filled with common quartz. **Chrysocollic chalcedony,** which occurs in the copper-ore mines of Arizona, is considered to be the most beautiful of all chalcedony varieties. This is a chalcedony strongly coloured by the blue-green chrysocolla, which sometimes even forms pseudomorphs after azurite crystals. Specimens of this particular mineral are much sought by collectors and are highly prized. Beautiful kidney-shaped coverings of chalcedony, in a lovely pale reddish-grey colour also develop in cavities of volcanic rocks near Chihuahua in Mexico. These cavities are also noted for containing perfect pseudomorphs of chalcedony after

153 Chalcedony — chrysoprase — Ząbkowice (Poland); 14 × 8 cm.

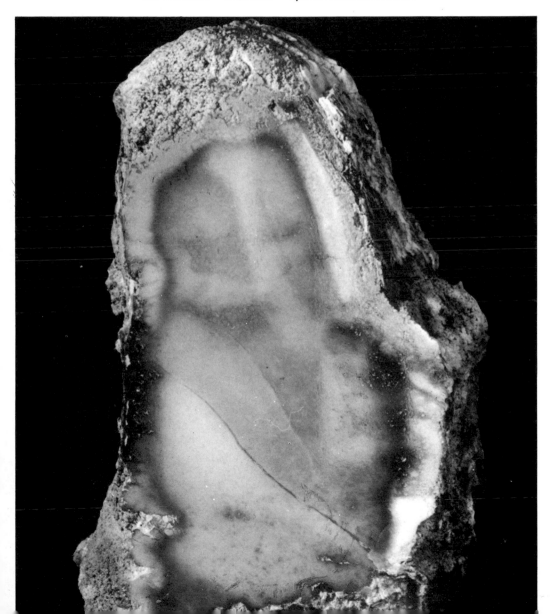

154 Chalcedony — moss agate — Železnice (Czechoslovakia); 9 cm.

155 Chalcedony — moss agate — detail.

156 Chalcedony — moss agate — Collyer (Kansas, USA); actual size of illustrated section 12 × 9 cm.

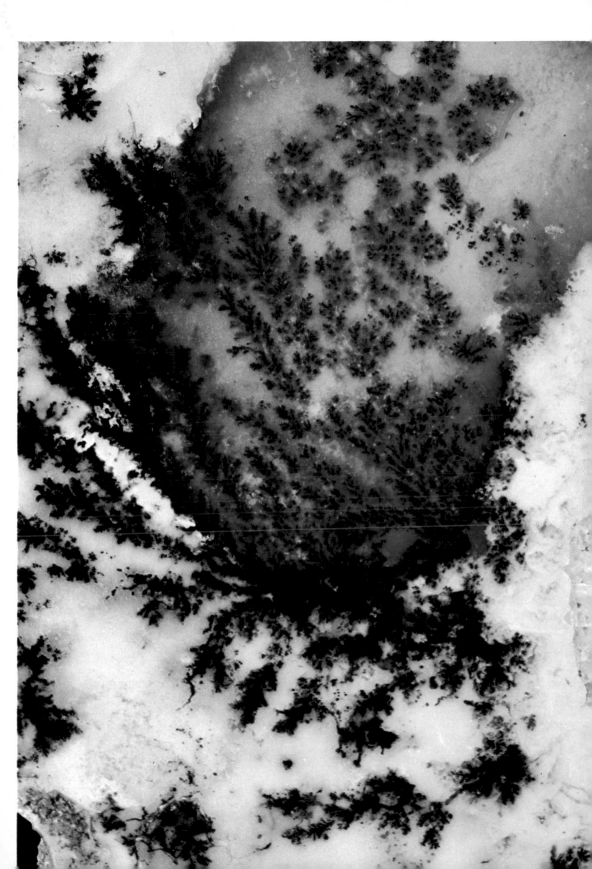

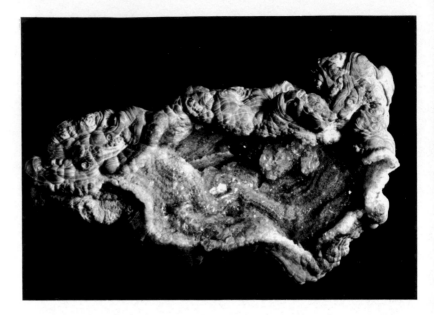

157 Chalcedony — enhydros — Uruguay; 25 × 13 cm.

tabular pseudohexagonal intergrowths of aragonite, which are sometimes as much as 8 cm in size. Chalcedony also occurs widely in volcanic rocks in Panama.

Chalcedony, like jasper and agate, is used in the manufacture of a great variety of ornamental objects. Chalcedony and its varieties are very popular with collectors. Because of its hardness and chemical resistance, which differs only slightly from quartz, massive pieces of chalcedony serve as an excellent raw material from which to make bearings and edges for sensitive scales, compasses and other instruments, pestles and mortars for chemists. A finely ground chalcedony is used for polishing many hard substances and for filling some soaps. Powdered chalcedony is also added to dyes and paints.

Chalcedony as a precious and a decorative stone enjoys popularity even today, for it is a constant, and a comparatively cheap, raw material for the manufacture of jewelry and small objects of art. The physical and chemical properties, including its hardness, which is only slightly lower than the hardness of quartz, are also very favourable when using chalcedony for jewelry. But chalcedony's popularity rests chiefly in the wide range of its often vivid colours.

**Agate** (158—169) is and always has been one of the most popular precious stones. Agate was first mentioned in writing by the Greek philosopher, surgeon and natural scientist Theophrastus (376—287 BC), but of course agate was known and used long before that. Theophrastus gave the stone its name — apparently because it was first discovered by the river Achates (now called the Dirillo) in southern Sicily.

Silicon dioxide, SiO₂, mixture of cryptocrystalline chalcedony, rhombohedral quartz and shapeless opal. H. approx. 7; Sp. gr. fluctuates; colour – see text; S. white to grey.

As agate is so conspicuous and brightly coloured, it is understandable that it was one of the earliest stones to be noticed. It was already used by the ancient Sumerians and Egyptians. Practically every culture of ancient times used agate not only for ornamental purposes and for making a variety of vessels, but also for making all sorts of amulets. The famous cameos and gems were carved from agate and so were beautifully engraved decorative articles, which were highly valued. Agates are most suitable for this purpose, for their variance of colour and unrepeatable pattern ensure originality when used as a precious stone. The earliest agate gems usually represented various religious symbols. As the occurrence of these most beautiful natural stones is so very rare, it is no wonder their presence was considered something extraordinary, and even that it was caused by supernatural powers.

As the standard of working the agates was very high in ancient Greece, articles made from them came more and more to the notice of the outside world. This raw material was turned into the most magnificent works of art — creations of absolute perfection and accuracy. The names of the Greek artists (for example Diodoras, Semon, Daidolas and others) have survived throughout the ages. The most famous was Pyrgoteles, the stone engraver of Alexander the Great; ancient and modern forgers have copied his monogram. Apart from such perfect works of art, less demanding articles, such as seals and ornaments, were made from this mineral.

The Romans, inheritors of the Greek culture, soon learned the art of working agates, especially how to make the traditional cut seal rings. They also gained a great deal of knowledge from the cultures of Asia and Africa, with which they were in regular touch. The oldest Roman agate gems, formed as scarabs, are a proof of this. Agate rings were worn in Rome from very early days. But originally only the Roman patricians were permitted to wear them. Hannibal, after his victorious battle at Cannae (216 BC), where he conquered the Roman legions, was able to determine the number of dead leaders and their rank, from their rings bearing the rank insignia. He sent the agate rings to Carthage as a war trophy. Round the year 85 BC the very first collection of agate gems was founded in Rome, the so-called dactylioglyphs. But a much more valuable collection of far greater beauty was based on the war spoils from the battles between Pompey and Mithridates. In 61 BC it was exhibited at the Capitol as a thanksgiving to the gods for victory.

After the fall of the Roman Empire, the traditional art of working the agate moved east to Byzantium. But the exquisite works of this era, engraved in agate, must be attributed mainly to Greek artists. The art of colouring agates black or red by heating was very popular and was a speciality of craftsmen in Constantinople. These methods were kept secret within the stone engravers' circles, and were passed from generation to generation through many centuries. Even today we do not know some of the procedures used by the artists of that era. After the fall of Constantinople and the Turkish occupation, the glyptic art was on the decline, and so was the popularity of agates.

Agates usually occur in amygdaloidal holes of melaphyres and similar rocks. Such rocks have originated through volcanic activity. After the gases escaped from the lava further, numerous crevices and cavities were filled with concretions of agates, formed by the cooling of rock solutions in them. Such fillers, which are much harder and more resilient than the rock itself, also occur in the topsoil when it is formed by weathering of the

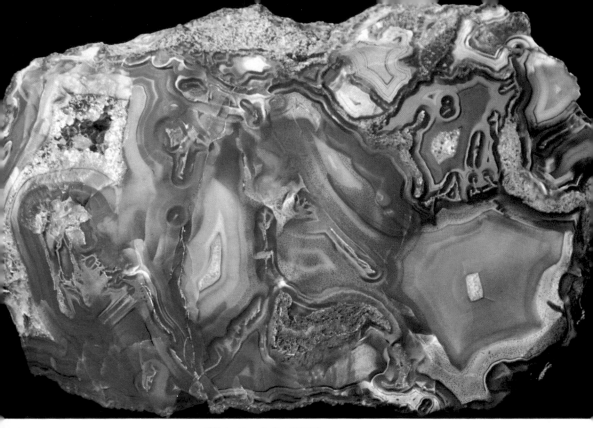

158  Agate — India; 28 × 17 cm.

159  Agate — Brazil; 10 × 6.5 cm

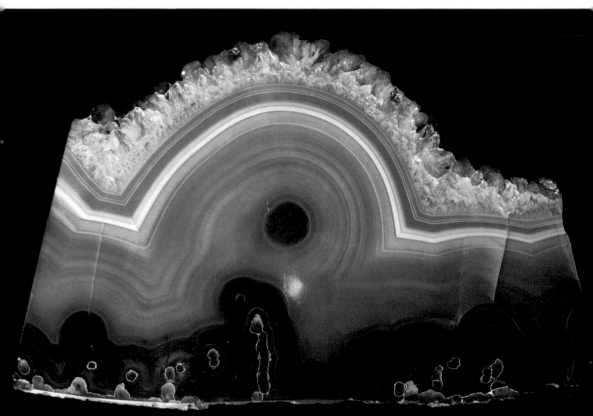

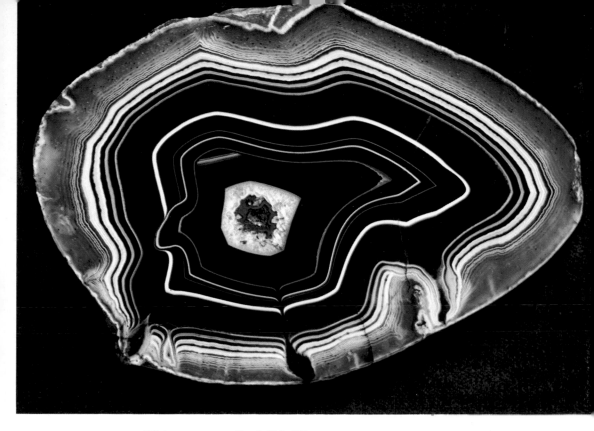

160 Agate — onyx — Brazil; 10.5 × 7.5 cm.

original rock. The agate layers are usually gaily coloured and as a rule the individual bands mirror the shape of the cavity in which the original siliceous solution mixed. The systematically arranged bands are clearly visible once the agate is cut and polished; in fact sometimes the opening, through which the solutions and all the colouring additions entered, is also distinguishable. In the centre of agates which have not been completely filled, crystals of rock quartz, smoky quartz or amethysts frequently occur. Less commonly the bands are developed in the form of wide ribbons. Agates have sometimes been given various additional names according to their colour, or the pattern and structure of their design. We distinguish between cloud-shaped agate, ogle-eyed agate; some are called star-shaped, coral, ribbon-like, spotted, agate of the ruins, etc. There are many more types. Agate which has white and black bands was given the proper name of **onyx**. This title, which in Greek means 'fingernail', was originally applied by Theophrastus and Pliny to various two-coloured and three-coloured layered stones. **Sardonyx**, named after the ancient city of Sardis, capital of Lydia (now in Turkey), which is a centre for trade in the mineral, is an agate that has alternating bands in orange-brown to red-brown and in white. It is also known as **carneolonyx**. The banding gives agates an unusually interesting and unique character. These variegated bands are formed by alternating layers of chalcedony and quartz, or of opal. The intricate structure of the seemingly compact

151

agates is best visible when thin sections are examined under a microscope. Depending on the minerals of which the layers are composed, they can be finely or more coarsely fibrous, granular or compact. The added mineral colouring is mainly due to the finely scattered powdery haematite, limonite and the greenish chlorites. The particularly vivid coloration displays a great variety of shades. Between the finely layered agates and the compact chalcedony many intermediate stages can be observed; chalcedony, after all, does not differ even in colour from its own agates.

Some agates are finely porous and so can be artificially coloured. Brazilian agates are highly suitable for this, for the porosity of their individual layers differs greatly; the less porous layers do not take colour and remain white or naturally pale. This variation was taken advantage of early for, as has already been mentioned, Constantinople in Byzantium already knew how to colour agate by artificial means.

With artificial dyes it is possible to achieve a wealth of shades. The oldest way was to colour the stones black, which resulted in the artificially coloured black and white layered onyx. The process was very simple. Selected agates were immersed in warm honey or in sugar solution for several weeks. The honey or the sugar gradually penetrated into the porous layers, till they were totally impregnated. By submerging the agates next in sulphuric acid, the organic substance was altered through a chemical process into black carbon. The latter stayed firmly embedded in the pores of the stone. The less porous layers, which rejected the organic substances, remained white. Today inorganic dyes are mostly used in adding colouring to agates.

161 Agate — onyx — Uruguay; 7.5 × 6 cm.

162 Agate — Brazil; 12 × 7 cm.

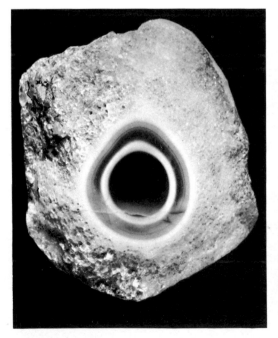
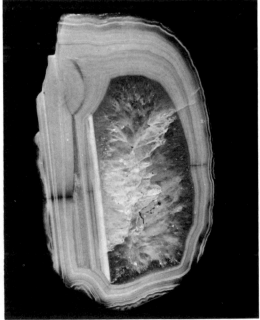

163 Agate — foothills of the Krkonoše Mountains (Czechoslovakia); 10.5 × 7.5 cm.

164 Agate — foothills of the Krkonoše Mountains (Czechoslovakia); 12 × 6 cm.

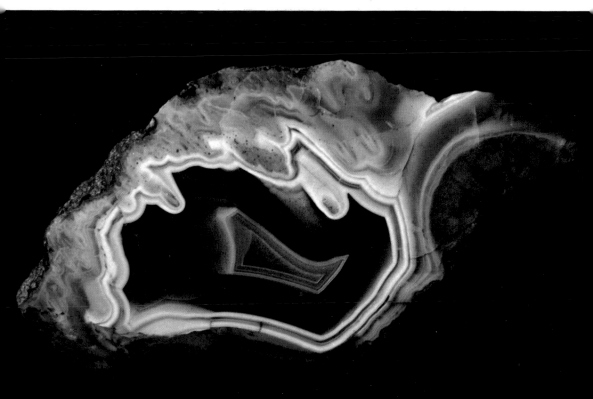

The most noted agate deposits are in Brazil, especially in the state of Rio Grande do Sul near the Uruguay border. Agates found in Kathiawar, India, in China, in the Urals, in Siberia and in Iceland are not of such high quality. The best European deposits are at Idar-Oberstein in the Rhineland (though they are now practically exhausted), in the foothills of the Krkonoše Mountains of Bohemia, in Saxony, mainly at Schlott-witz, and in Romania.

During the 14th century agates were recovered mostly from numerous sites in Saxony, and during the 16th and 17th centuries from the foothills of the Bohemian Krkonoše Mountains. The variability of the banded designs of agates once again commanded attention. They were again selected as the most apt material for the manufacture of ornamental objects, often with beautiful engravings. Such objects were extremely expensive, for they were the works of art of the most noted artists of that period. They were often moulded according to antique models which were still preserved, and which were bought at fantastic prices and amassed in collections of rulers and the gentry of the Renaissance period. The largest collection of this type was started at the Prague court of the

165  Agate — Tietê (Brazil); 9 × 6.5 cm.

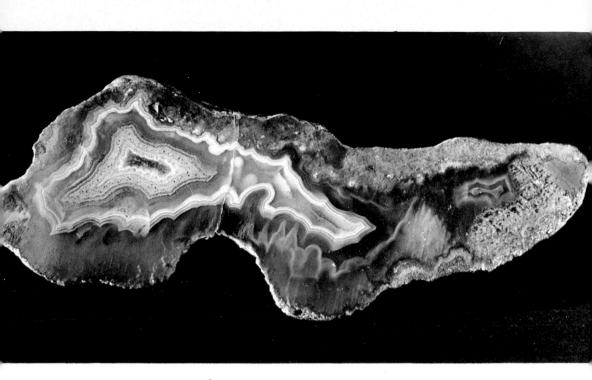

166 Agate — Železnice (Czechoslovakia); 12 × 5 cm.

Emperor Rudolf II. One of the items was a sardonyx with an engraving which represented the triumph of Augustus' step-grandson Germanicus, an antique gem called *Gemma Augustea*, which he bought for 12,000 ducats. Today this most famous example of antique glyptics is in the collection of the Museum of Applied Art in Vienna. Of all the magnificent creations of the Renaissance artists from the court of Rudolf II only a few agate urns are preserved; these today are a part of the Viennese court collections.

The great expansion in the industry and the period when agate products gained high popularity began in the Modern Age after 1827, when the agate and chalcedony deposits in Brazil were first exploited. This happened just when the deposits of agates and chalcedony in Idar were almost exhausted, due to the enormous industrial expansion, and the grinding plants were being threatened with a shortage of the raw material. The stones imported from the abundant deposits of India were extremely expensive and often of inferior quality. Many of the grinders and engravers, fearing that an inevitable crisis was approaching, decided to emigrate to South America, particularly to Brazil, which by then was renowned for the wealth of its extensive deposits of precious gems. Some of the immigrants settled in the Rio Grande do Sul state, which borders on Uruguay; and here in 1827, they found rich deposits of agates and chalcedony, which were being rather lavishly and wastefully used locally as paving stones. For transporting such valuable material into Europe the

155

167 Agate — Horní Halže (Czechoslovakia); 12 × 9.5 cm.

168 Agate of the ruins — Schlottwitz (German Democratic Republic); actual size of the illustrated section 5 × 5 cm.

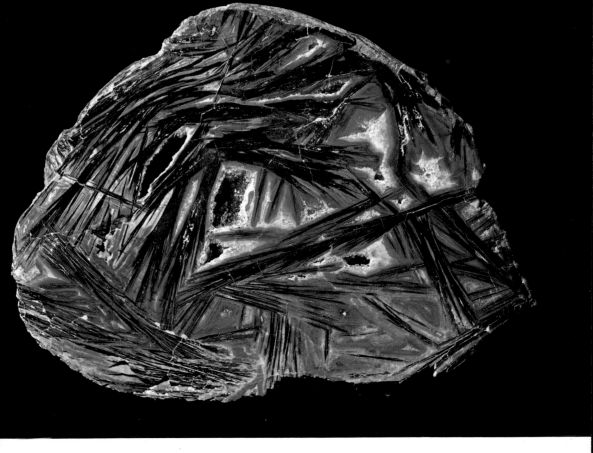

169 Agate — Trestia (Romania); 25 × 19 cm.

immigrants employed empty ships, which willingly took as ballast for only a small fee or for no fee at all, barrels and sacks of animal skins, filled with agate and chalcedony nodules. The Idar industry thus gained fresh, cheap and excellent raw material, and unprecedented prosperity. The most noted collection of our time of antique gems and objects made of agates is in the Bibliothèque Nationale in Paris.

Although the popularity of agates and of jewelry and objects made from them is not as great as in the past, they are still used for the manufacture of art objects, especially dishes, goblets, boxes, writing-sets, paperweights, ashtrays, etc. Seal rings are still popular, especially those made from onyx and carneolonyx (sardonyx).

Agates are also widely used in industry. Because of their hardness and resistancy, massive pieces of these minerals make an excellent raw material for the manufacture of bearings and edges of accurate scales, compasses and other instruments, and mortars for chemical purposes.

**Opal** (170—183) is a mineral whose colour, or rather the iridescent play of colours, is practically unique and impossible to imitate. It is natural then that it was one of the first precious stones to be used for ornamental purposes and it was highly valued even in ancient times. The earliest extant objects made of opal date back to approximately 500 BC.

Hydrous silicon dioxide $SiO_2 +$ $nH_2O$, amorphous. H. 5.5–6.5; Sp. gr. 1.9–2.3; colourless, white, yellow-red (fire opal), rarely differently coloured, often displaying opalescence, with waxy, greasy or pearly lustre; S. white to colourless.

Its name is an old Latin term, used by Pliny the Elder, at first for only the precious opal. Pliny writes: 'The flat precious stone called opalus is the most valuable of all the stones, but it is difficult to define it and describe it. It has the gentler fire of the ruby, the brilliant purple of the amethyst and the sea-green of the emerald, all shining together in an indescribable union.'

It is obvious that even then pastel shades of opal were greatly valued — the blue-green, the blue-violet, and the fiery red in the most delicate hues. Long before Pliny, as far back as the 6th century BC, the mystical Greek poet Onomakritos sang the praises of the beautiful internal play of colours, and Dioscorides, the Greek surgeon and naturalist (c. AD 50) classed opal as one of the most valuable gems.

It is said that the Roman senator Nonius, who was the owner of a fine and valuable engraved opal, chose exile rather than surrender the stone, which was no bigger than a hazelnut, to the great commander Marcus Antonius.

The popularity of opals, especially of precious opals, and those with an exceptional play of colours, has lasted through the ages and continues today. The most famous historical opal was the so-called 'Trojan Fire' from the French crown jewels, which belonged to the French Empress Josephine. It was lost during the French Revolution.

Opal forms shapeless, amorphous stones of reniform, stalactite or botryoidal shape. Frequently it fills cavities of rocks. It also occurs in veins and bands, or rarely as pseudomorphs after other minerals. The colour differs according to the variety. The content of water, which is not constant, has a great influence on the play of colours. Opal usually contains 3 to 13 per cent water, but some contains as much as 34 per cent. On the other hand some opals are practically without water.

Opal aggregates are compact. Frequently interstages and mixtures of opal, chalcedony and quartz occur. According to the coloration, transparency and occasionally the structure, a number of interesting opal varieties are distinguished. The best known, most sought-after and most valuable variety is **precious opal** (171—173), a stone with the most beautiful and radiant, yet uncommonly delicate colours. When pure it is usually milky-white or faintly yellowish. Sometimes it has also coloured stripes.

The precious opal is outstanding among all the precious stones for it has the most perfect, brilliant iridescent play of colours. This is caused through the breakdown of light rays when they meet finely scattered foreign particles, or minute water-filled cavities, or round clusters of amorphous silicon dioxide, as has recently been proved with an electron microscope. When expertly treated, this optical characteristic of opal can be emphasized even more. Opal therefore is considered one of the most valuable precious stones, though it has the lowest hardness of them all (only 5 to 6.5 on Mohs' scale). In spite of this deficiency, which would have been unforgivable in other stones, opal rouses great enthusiasm.

**Black opal** is the most valuable of all opals; it is a dark variety of precious opal with an intensive colour-change, and was first found at the end of the 19th century in Australia. For a while black opal was valued as highly as diamond. **Fire opal** (175—176) is the most beautiful opal variety; it is richly red like a hyacinth and transparent, with excellent opalescence, especially after being polished. The colouring fluctuates between light yellow-brown and rich brownish-red. The stone was brought into Europe from Mexico by a German natural historian, Alexander von Humboldt (1769-1859). He found it while studying local volcanic activities.

170 Common opal — Herlany (Czechoslovakia); 11 × 7 cm.

**Hydrophane** is mineralogically most interesting. It is a precious opal which, when exposed to air, loses water and therefore also the characteristic play of colours. When immersed in water, the play of colours returns. From a jeweller's point of view, valuing such a stone is most problematical.

The play of colours is characteristic of only some of the opal varieties. The other opal varieties which lack the internal play of colours, are very rarely used as ornamental stones. The clearest, purest opal, **hyalite** or **glassy opal**, is completely clear. The title is derived from the Greek 'hyalos' — glass. Usually it occurs as finely botryoidal or stalactitic aggregates with a reniform surface and displays a strongly vitreous, though somewhat greasy lustre. Its cold, icy beauty is strangely impressive. The most widely known is the **semi-opal**, which is compact, non-trasparent to translucent, and occurs in various, often bright colours. **Common opal** (170) has different names according to its colouring.

159

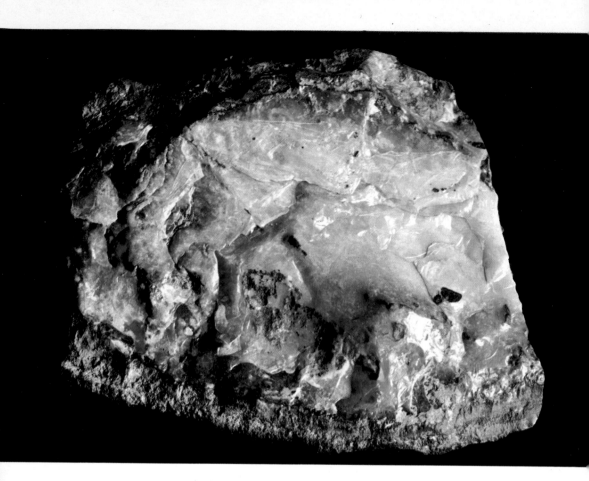

171 Precious opal — Barrace River (Queensland, Australia); 10 cm.

It is commonly compact and non-trasparent, cloudy, in a variety of colours, especially in the darker shades. Often it is even vividly coloured. Generally it is found in certain silicates, particularly in serpentines, which decompose to form the irregular, sometimes strikingly beautiful tubers of opal.

**Milk opal** is milky white, and is a mixture of opal and crystalline aggregates of silicon dioxide. Its faces are usually translucent. It if happens to be interwoven with a network of fine black dendrites of hydrous oxides of manganese, it is called **dendritic opal**. Such a mineral gives the impression of containing tufts of moss, which display such regularity, perfection of branching and minutest details that they are an impressive image of plants turned to stone.

The colouring of opals is caused by the presence of oxides containing

172 Cut precious opal — Dubník (Czechoslovakia); 28.9 × 21.7 mm; weight 22.64 carat.

173 Precious opal — Dubník (Czechoslovakia); 9 cm.

metals, especially iron (**wax opal** for instance), and also by the presence of foreign bodies (inclusions). **Serpentine opal** belongs to this group. Green in colour, it contains a multitude of tiny serpentine particles, and was formed through the decomposition of serpentine. **Forcherite** is coloured by orpiment to orange-yellow, and is an opal found in cracks of gneisses at Knittelfeld in Styria (Austria). The very similar yellow-brown **fiorite** forms reniform concretions in trachyte cracks at Santa Fiora in Tuscany, northern Italy. **Jaspopals** are richly and gaily coloured mixtures of opal and chalcedony, microcrystalline quartz and mineral pigments. **Prasopal** is coloured green by aqueous nickel metasilicates. It occurs at Koźmice in Polish Silesia and also in New Caledonia.

**Menilite** (177) from the Ménil Montant deposit in Paris is a grey-brown, distinctively layered opal concretion resembling a claw. **Cacholong** (174) is a chalky-white mixture of opal and chalcedony. **Wood opal** (180–181) is formed by opal permeating through wood matter, but retaining the wood structure. One of its varieties is **tabasheer** permeated into bamboo stalks.

Opal is by no means as abundant in nature as quartz. Generally it resulted as a gel from solutions. It occurs through the deposition of hot solutions, from hot springs (geysers) or through the decomposition of silicates in rocks, especially in serpentines. It develops similarly in andesites and basalts. Opal rocks, tripolites (diatom slates), named after occurrences in Tripoli, North Africa, originated from deposits of many minute silicified skeletons of algae, called diatoms. They are usually earthy in appearance. If they are of a loose consistency, they are called diatomaceous earth, or infusorial earth (the name is incorrect, but earlier natural scientists used to think that this material was composed of the skeletons of infusorians). Cacholong and menilite are of organic origin. They occur therefore in sedimentary rocks. Opal is also found in sandstones.

The oldest deposits of precious opal are at Dubník near Prešov in Czechoslovakia, known all over the world for its 'Hungarian opal'. Apparently the deposit was known in ancient times, and many of the world's best known precious opals were recovered there, including the 'Trojan Fire' mentioned earlier. The mining of opals was carried on there until this century. Even in 1920 a high quantity of the raw material was extracted from the Dubník mines, from which it was possible to grind 1,600 carats of opals of an exceptional gem quality. It is still possible to find small fragments of precious opal on pit heaps. Opal occurs in these deposits in the cracks and crevices of andesite rocks, especially breccias and tuff. It is particularly noticeable for the beautiful, soft play of colours — violet-blue, blue-green and red hues.

Other varieties occur with the precious opal, such as the crystal-clear **hyalite** (182–183), milk opal, hydrophane and various common opals. The volcanic rocks from the Tertiary Period and their tuffs, especially the andesite and trachyte ranges, in the whole region of central and southern Slovakia yield a wealth of deposits of common opals.

The largest and the most beautiful opal preserved to this day was found in the bed of a stream in 1775 in Dubník. Today it is in the collection of the Natural Science Museum in Vienna. It weighs approximately 600 grammes, and measures $12.5 \times 5.7$ cm. The largest collection of opals from the Dubník deposits is in the Natural Science Museum in Budapest. Altogether it contains 366,926 pieces with a total weight of 58,645 carats. The discovery of Australian and later Mexican opals of more vivid colours, brought the glory of the Slovakian opals to an end. Another cause of the decline was certainly the fact that the mining there was relatively unproductive in the last few years as far as good-quality stones were concerned; the modern superstition, that opal is a stone which brings bad fortune also did not help. It is indeed strange how such

174 Opal — cacholong — Olomučany (Czechoslovakia); 14 × 12 cm.

a beautiful stone could fall into disfavour in such a short space of time.

The richest contemporary deposits of opals are in Australia, in New South Wales, Victoria and Queensland. These deposits were discovered in the second half of the 19th century, and an interesting tale is spread about their discovery. According to the story, a hunter shot a kangaroo in the outback; the animal fell to the ground, but kicked the earth with such force that soil flew to all sides. Suddenly the soil glowed with vivid colours. The hunter filled his pockets with chippings of a strange stone. In Adelaide he showed them to a goldsmith, who immediately recognized that they were the precious opal. He bought up all the pieces — and this was the start of 'opal fever'. The outback filled with hopeful opal searchers, and they founded the small town of White Cliffs. According to historical sources, Australian opals were first recovered in 1877.

In the Australian deposits precious opals occur in many varied forms. The stones which stand out with their exceptionally intensive play of colours are called 'harlequins'. The most valuable black opal occurs only in the deposit at White Cliffs in New South Wales. From here come also the magnificent pseudomorphs of opal after the crystals of the mineral glauberite, a sodium and calcium sulphate, and perfect opal inclusions of shells, bones and other organic remains. Opal-minig is of great economic significance in Australia.

The biggest uncut Australian opal is in the possession of the Natural Science Museum of New York. The biggest cut and polished Australian opal is at the Smithsonian Institution, Washington and weighs 155 carats. Towards the end of the 19th century yet another outstanding stone was discovered in Queensland. While it was being recovered, it split in half. One piece is now among the Royal treasures of Britain and weighs 250 carats. The colours are very rich — scarlet, amethyst-violet, dark-green — with a wondrous play of colours. There are other deposits

175 Fire opal — Zimapán (Mexico); 9 × 7.5 cm.

176 Cut fire opals — Zimapán (Mexico); largest oval piece 19 × 15 mm; weight 11.285 carat.

of precious opal in Virgin Valley, Nevada and in Mexico, but they are not as important. New, promising deposits of opals have been discovered in the state of Idaho, Washington and Oregon (USA), where the opals occur as small tubers in volcanic rocks. The Mexican deposits though they have not the same significance as the Australian deposits, are scattered over a gigantic area of volcanic rocks from San Luis Potosí right up to Guerrero. The most beautiful specimens are extracted from the Guerrero region and recently also from the state of Jalisco. As with stones from Australia, the most gaily coloured Mexican opals are called harlequins and are noted for their exceptionally intensive play of colours. This jewelry term is also applied to the precious opals, where variously coloured individual grains display opalescence. Other deposits of precious opal are in several locations in Honduras and in limited quantities also in Brazil, again in several localities. Here the precious opal is usually accompanied by numerous varieties of common opal and fire opal, which has recently been discovered also in the USSR. Northern Mexico has deposits of beautiful fire opals. They have been recovered there since 1870, and form in cracks and crevices of porphyrite trachyte. Similar fire opals also occur in Asia Minor, but they have not as yet been extensively mined. In India there are many occurrences of different varieties of common opal (dendrites). Wood opal occurs in the state of Idaho, USA, where, in the deposits of Clover Creek, beautiful trunks of oak are found permeated with opal. The USA is exceptionally rich in similar

165

trunks of different types and different ages. It is necessary to mention at least one such deposit in layers of volcanic ash in the central regions of Washington state and in similar layers of ash and tuffs in the deposits of Virgin Valley, Nevada. The local opals, which fossilize wood, display an intense play of colours, fiery red, green, blue and violet, which penetrates through the dark brown to black tree-trunks. But the deposit is noted also for the pale to white trunks. These lumps of wood, permeated with precious opal, frequently weigh as much as 3 kg. Other known occurrences of trunks penetrated by opal are in Oregon and Utah. The biggest trunks of this kind are found, however, on the Australian island of Tasmania, in the Bothwell deposit.

The now exhausted deposits of hyalite in the basalt rocks of Valeč in

177 Opal — menilite — Mokřina (Czechoslovakia); 12 × 9 cm.

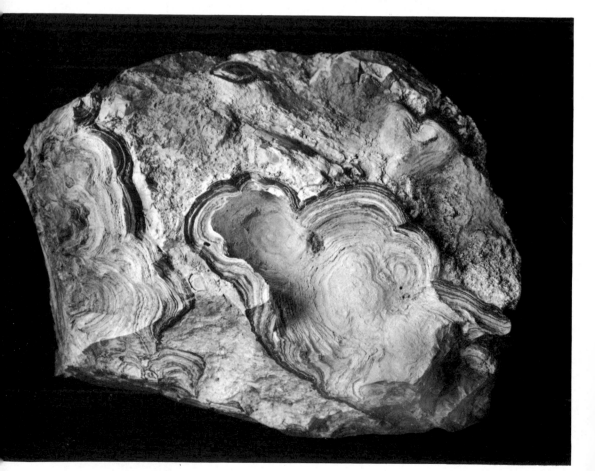

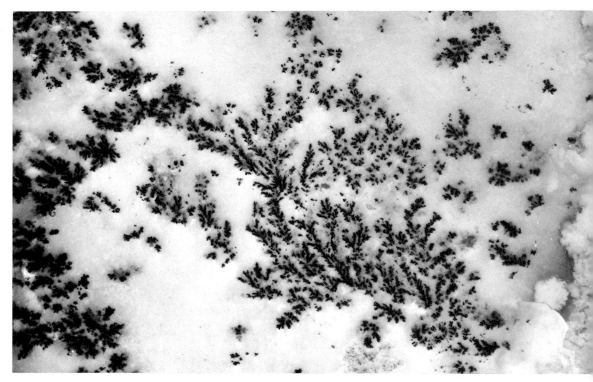

178 Dendritic opal — India; 11 × 8 cm.

179 Diatomite — Bechlejovice (Czechoslovakia); 14 × 8 cm.

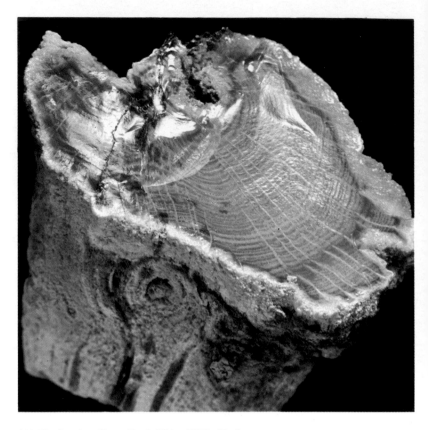

180  Wood opal — Clover Creek (Idaho, USA); 15 × 9 cm.

Bohemia are well known. Hyalite is now recovered mostly from Mexico and Potosí in Bolivia. It also occurs in the form of tiny spheres in Japan. These are formed by hot solutions permeating the cracks and crevices of rocks during dying volcanic action, leaching them and then depositing grape-like clusters of unusual beauty in cooler places near the earth's surface.

The deposit Cerro del Tepozan near San Luis Potosí is very popular with collectors from all over the world. Hyalite forms here thick crusty covers in cavities of volcanic rocks (liparites). The occurrences of hyalite in basalt near Klamath Falls in Oregon, USA, are very similar. The hyalites from New England and also from North Carolina, USA, are famous for their intense green luminiscence in ultraviolet light, which apparently is caused by the admixture of collodial fragments of uranium minerals. These luminescent hyalites form thin crusts on cracks of pegmatites.

Apart from Ménil Montant (Paris), menilite also occurs in western Bohemia, and cacholong occurs in Mongolia. Opal's variety, **geyserite** forms from deposits of springs around geysers in Iceland and in Yellowstone National Park in the USA.

Opals which have an attractive play of colours are much sought after.

181 Wood opal — Brhlovce (Czechoslovakia); actual size of illustrated section 9 × 6 cm.

182 Opal — hyalite — Valeč (Czechoslovakia); actual size of illustrated section 6.5 × 5.5 cm.

Semi-opals, too, are often used for ornamental purposes. Dendritic opals are most suitable for this. Cutting and polishing enhances and magnifies their branching, interwoven patterns. Wood opals are also suitable.

Otherwise opals are usually cut in the cabochon cut; the cut with facets is applied only very exceptionally (to fire opals of the highest quality for instance). When cutting precious opal, or fire opal, more attention is paid to enhancing their iridescence than to the shape of the actual cut. The low degree of hardness makes opals unsuitable for use anywhere where there is a danger of scratching the stone or of other mechanical damage.

In 1921 the art of colouring opals by synthetic means was discovered completely by chance. Bittner-Belangenay, a Frenchman working on the Dubník deposits, happened to spill ink on a few opals on his table. The ink coloured some of the stones so perfectly, that even when completely dry, and even after washing and cutting, they shone with unusual, unknown colours and hues. Opals coloured by such artificial means were given the name of 'chameleon'.

It was believed in ancient times that an opal brought its owner good luck. It is therefore difficult to explain the modern superstition that this beautiful and popular stone brings bad fortune, which is completely in contrast to the older belief. It is said the rumour was spread by the owners of the opal mines at Dubník in an effort to stop the extensive stealing of opal in the 1870s. It seems they were partially successful.

Diatomite is used as a metal polisher, in the manufacture of dynamite, as an insulator for high temperatures and in filtering chemicals. Geyserite is used in the manufacture of some ceramic products, such as special enamels. Diatomaceous earth is used far more extensively for technical purposes in the most varied branches of industry. It is particularly useful because it is highly porous and a poor conductor of sound, heat and electricity. In the loose earth form it is therefore valuable raw material for the manufacture of light bricks for furnaces, filter-material, cleaning pastes, abrasive stones, etc. It is also used in the manufacture of primary explosives, paper stock and varnishes, and as a filler of soaps. Diatomaceous earth is mined chiefly in Algeria, California and Canada; in Europe, the main producers are northern Germany, Tuscany (Italy,) Bohemia and southern Slovakia.

183 Opal — hyalite — San Luis Potosí (Mexico); globular aggregates up to 3.5 cm.

**Psilomelane** (184—185) was known to Theophrastus (372-287 BC) and Pliny the Elder (AD 23—79) who mentioned it as a 'non-magnetic' variety and specified it as 'magnesium' for discolouring glass. The manganese element was only discovered at the end of the 18th century, when the mineral was thoroughly tested. Psilomelane usually occurs in deposits in association with two closely related minerals, pyrolusite and manganite, and often in an earthy admixture of iron (wad). Psilomelane as well as the more common black-brown **wad** (185), is often found in a dendritic, tree-like form in crevices of various rocks, or in stalactite and stalagmite masses. The most important deposits of psilomelane and other manganese ores from the commercial point of view are in the Ukraine and in India. Psilomelane crystals are also found in Lostwithiel (Cornwall), England and Batesville (Arkansas) and Catersville (Georgia), USA. Most of the manganese mined is used to improve the quality of iron. In the production of steel for instance, ferro-manganese, an alloy of iron and manganese, is used. The presence of manganese in steel is responsible for its flexibility, cohesiveness and resistance against impacts. Manganese is also used in the chemical industry in the manufacture of disinfectants, paints, batteries and matches. Psilomelane, like similar oxides, contains over 60 per cent manganese.

Complex oxide of manganese and barium, $(Ba,H_2O)_2Mn_5O_{10}$, monoclinic. H. up to 6; Sp. gr. 3.2–4.7; black, submetallic lustre; S. brownish-black to black.

184 Psilomelane — Erzgebirge (Czechoslovakia); 8 × 5 cm.

185 Wad — dendrites — Pappenheim (Federal Republic of Germany); 12 × 8.5 cm.

186  Cassiterite — Horní Slavkov (Czechoslovakia); 16 × 12 cm.

187  Cassiterite — Cornwall (England); 9 × 9 cm.

**Cassiterite** (186 — 187) has served man well as a tin ore from ancient times, for it can be easily worked and extracted. Man had already learned to melt it by the 6th millennium BC, before he knew how to melt iron. Cassiterite was heated with charcoal in simple clay or stone kilns, where a temperature of 1000°C could be reached, and produced tin. As copper ores were often present with cassiterite and they were all melted together, an extremely hard yellow alloy of tin and copper frequently resulted. This is how men first came to manufacture bronze, a metal of far superior quality than the soft tin. Many useful objects were made from bronze.

Tin dioxide, $SnO_2$, tetragonal. H. 6–7; Sp. gr. 6.8–7.1; brown to black, more rarely yellow to reddish, with diamond to metallic lustre; S. pale brown to white.

Cassiterite forms perfect columnar crystals, often twinned. More often it is massive, granular, or in the form of fibrous aggregates. Sometimes it is disseminated in rocks (such as greisen), or it forms lodes, often in association with tungsten and molybdenum ores. It is also mined from alluvial placers. The largest alluvial deposits are in Malaysia (Kuala Lumpur), which yield one-third of the world's production, and in Indonesia (Bangka and Belitung Islands).

The largest primary deposits are in Bolivia. In Europe the chief ones are in Cornwall, Brittany and Bohemia, and in the United States at El Paso, Texas and the Black Hills of South Dakota. Tin is used today mainly in wrapping and preserving food (tins, tin foil), for it is a metal harmless to human health.

There are very few metals which have a practical significance and which are as poor in their own types of ores as is tin. Cassiterite happens to be almost the only ore, but a substantial one, found in large quantities in some localities. It contains 78.6 per cent tin. It is the very best ore of tin, for the remaining tin ores do not carry such a high content of metal and, as a rule, do not occur in sufficient quantities to warrant mining. It is therefore not surprising that the mining of cassiterite is on the rise in all deposits. Cassiterite is also a popular mineral with collectors. In particular its crystals, which usually occur as twins or multiple twins, are highly prized.

173

**Rutile** (188—191) has long been found in many alluvial placers and in quartz veins in the form of columnar crystals. Because of its appearance, it was often taken for tourmaline. In 1795 the German chemist M. Klaproth found the mineral contained a substantial amount of titanium. The present name was a suggestion of a German mineralogist and geologist Abraham Gottlob Werner and originates from the Latin 'rutilus' — golden-yellow, reddish-yellow, ginger. No one visualized in those days how important rutile would be as a titanium ore.

Titanium dioxide, $TiO_2$, tetragonal. H. 6; Sp. gr. 4.2–4.3; brownish-black, scarlet, more rarely yellow-brown, also black (nigrine); vitreous to adamantine lustre, nigrine submetallic; S. pale brown to yellowish, grey-black to green-black.

Long before this, rutile was known in a completely different form. There were many places in the Alps where magnificent clear quartz was found, containing slender, needle-like, often thickly interwoven crystals; they often formed an unusual, interesting type of netting and helped to enhance the otherwise rather uniform mineral (188). This combination is known as sagenitic quartz and was thought to be at first an independent mineral. They were widely used for stones, for rings and necklaces, often under the title of 'the hair of Venus' or 'arrows of love' (flèches d'amour). That the mineral concerned was rutile was discovered much later.

The columnar crystals of rutile are often twinned in the form of geniculate twins, or complex multiple twins. Rutile often occurs as cobbles, or in granular or acicular form. Rutile's jet-black variety, rich

188 Rutile — sagenite in clear quartz (rock crystal) — Modriach (Austria); 3 × 3 cm.

189 Rutile — the Alps; largest twin 13 mm.

in iron, is called **nigrine**. This variety is abundant in schists, in igneous rocks lacking in silicon dioxide, in syenites and diorites, and in quartz veins of gneisses and mica-schists. As it is resistant to weathering, rutile also occurs in some alluvials. There are some magnificent rutile crystals in Modriach in Styria; sagenites are recovered from the Tirol and there are rutile deposits in Switzerland (St Gotthard), Bohemia (near Soběslav) and southern Norway (they are extracted from Kragerö, which is rich in rutile crystals), as well as in the USSR (the Urals), in many places in the USA (Arkansas, Georgia, Virginia and Pennsylvania) and in Brazil.

Rutile is a substantial source of titanium used in steel production, so it is widely used in modern industry. With iron, rutile forms an alloy, ferrotitanium, which improves the quality of steel, especially its resistance to corrosion and heat, and raises its weldability. It is widely employed in the aerospace industry. Titanium white and other alloys are all well known. It is also employed for cutting-tools. Rutile contains 61 per cent titanium.

Natural rutiles are only rarely used as precious gems. Then they are usually cut in the brilliant cut which makes the most of the high lustre. Synthetic rutile is one of the most beautiful artificial precious stones; in contrast to the natural stone, it is transparent. The first crystals were manufactured in Czechoslovakia in 1942, by the Verneuil process. However, in 1943, it was necessary to discontinue the manufacture of synthetic rutile until the war was over, when production restarted. But during the war period, the synthesis of the monocrystals of rutile was performed in the United States.

The chief value of the synthetic rutiles rests in their lustre, which even exceeds the lustre of diamonds. It is caused by the high refractive index (rutile = 2.6—2.9, diamond = 2.4), and by dispersion (the ability to break down white light to a coloured spectrum) and by the reflection

175

190 Rutile — Soběslav (Czechoslovakia); intergrowths 3—5 cm.

factor. These properties point to the possibility of establishing rutile as an important stone in the gem and optical industries. Its electrical properties are also promising for technical purposes. It has one great disadvantage, particularly in jewelry, which is the low degree of hardness. American rutiles, manufactured in great numbers since 1948, are marketed under the title of 'Titania' and are considered popular and most unusual gemstones. When suitably cut and polished, their exceptional optical properties really come to life. The yellowish tinge remains a disadvantage — so far no one has managed to remove it completely. On the other hand, firms in the United States have manufactured brown, orange and blue rutiles.

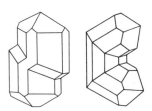

191 Rutile — sagenite — Diamantina (Brazil); 6 × 5 cm

**Anatase** (192) was given the name from the Greek 'anatasis' — elongation, because of the crystal form. It shares rutile's chemical composition and crystal symmetry, but has a different internal structure. The usual form of anatase crystals is of steep double pyramids with a particularly strong lustre. Anatase is far rarer than rutile. Most frequently it occurs in druses of metamorphic rocks and in veins of hydrothermal origin; brookite is often its associate in the cracks of gneisses. Main deposits: the Alps, Rauris near Bischofshofen in Austria, the surroundings of Kutná Hora in Bohemia, Minas Gerais in Brazil, and the Urals. Anatase has no practical significance, but is sought after by collectors. The transparent crystals of gem quality, found in the diamond-bearings gravels in Minas Gerais state, Brazil, are particularly valued. Beautiful blue anatase crystals have been recently discovered in diorites at Beaver Creek (Gunnison County), Colorado, USA. In the classical deposits in the valleys of the Swiss Alps, such as the St Gotthard Pass, new discoveries are constantly being made, which also applies to the deposit at Le Bourg d'Oisans (Isère, France) and to a number of deposits in the Italian Alps.

Titanium dioxide, $TiO_2$, tetragonal. H. 5.5–6; Sp. gr. 3.90; blue-black, black, brown, yellow, with adamantine or metallic lustre; S. colourless to pale yellow.

192 Anatase — Sonnblick (Austria); crystal $12 \times 9$ cm.

193 Brookite — Amsteg (Switzerland); crystals 6—14 mm.

**Brookite** (193) was so named in honour of the British crystallographer and mineralogist H. J. Brook (1771—1857). It shares the chemical composition of the two preceding minerals, but in contrast to them crystallizes in orthorhombic form. Crystals may be small and tabular. It always occurs in crystallized form. On rare occasions it occurs in gneisses, and even more rarely is found in druses of pegmatites (especially albite rocks). Main deposits are at Prägraten (Tirol, Austria), Switzerland, Miass (Urals, USSR) and Magnet Cove in Arkansas (**arkansite** variety). Collectors are particularly keen on the deposit in the Maderaner Tal (valley), Switzerland, where it is possible to find beautiful brown tabular crystals up to 5 cm in size, and also on the arkansite deposit, whose pseudohexagonal pyramidal crystals reach up to 3 cm. They are commonly embedded in cloudy-grey columns of quartz.

Titanium dioxide, $TiO_2$, orthorhombic. H. 5.5—6; Sp. gr. 4.14; yellow-brown to black, with adamantine to metallic lustre; S. colourless.

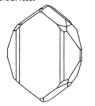

179

194 Tantalite — Bodenmais (Federal Republic of Germany); 3 cm.

195 Wolframite — Cínovec (Czechoslovakia); 6 × 5 cm.

**Tantalite** (194) occurs comparatively rarely as prismatic or tabular crystals, formed of several prisms, in pegmatites. As practically the only source of tantalum it is extracted only from secondary deposits and alluvial deposits. The main producers are Australia, Zaïre and, in Europe, Bodenmais in Bavaria (West Germany). Tantalite is used in the production of chemical instruments and surgical instruments, also in electrodes.

Oxide of tantalum and iron with niobium and manganese, $(Fe, Mn)(Ta, Nb)_2O_6$, ortho-rhombic. H. 6; Sp. gr. 7.9; black, submetallic lustre; S. black to brown.

**Wolframite** (195) was once considered as unwelcome admixture in melting tin ores. Apparently wolframite was given its name by old German miners, since during the process of melting it 'gnawed' (like a wolf) through cassiterite. The element wolfram, discovered in 1781, was named after wolframite. It is now more often called tungsten. For a long time tungsten was considered a valueless gangue material, though today it is sought after. According to the old grouping system, wolframite was thought to be wolframate (or tungstate) and was placed in the sulphate group. It was classified as an oxide comparatively recently. Columnar to tabular crystals, or crystals with perfect cleavage, and scaly or fibrous aggregates of wolframite occur most often in cassiterite veins, which they frequently dominate. The most important deposits are in the USA (Boulder County, Colorado, California and New Mexico), in Bolivia and South Korea; in Europe, in Portugal, England (Cornwall) and in the Ore Mountains, Bohemia. Tungsten is predominantly used in the steel industry in the manufacture of steels for high-speed tools and for the manufacture of filaments of electric bulbs, which sometimes have a diameter of as little as 0.01 mm. These uses depend on the exceptionally high melting point of tungsten (3,387°C — the highest of all metals). Many compounds of tungsten are excellent dyes and are also used for impregnation purposes in fire-proofing cloth. Compounds of tungsten are also widely employed in a number of other branches of industry. Wolframite is the most important ore of tungsten.

Oxide of tungsten and manganese with iron, $(Mn,Fe)WO_4$, monoclinic. H. 4.5–5.5; Sp. gr. 7–7.5; iron-black with a brownish shadow, submetallic to metallic lustre; S. black.

180

**Uraninite** (196 — 198) was mined at Jáchymov in Bohemia in fairly substantial quantities probably right from the onset of silver mining, and was at first thought to be an ore with a small admixture of silver. Later it was proved there was no silver present, so this mineral, which closely resembles pitch, was called pitch-blende and because it seemed to have no use, it was thrown away with other gangue. Today it is difficult to comprehend how the most valuable ore of such important uranium and radium compounds could have been thus discarded as useless material during the mining of other ores. Yet this is exactly what happened.

Uranium dioxide with thorium, $(U,Th)O_2$, cubic. H. 5–6; Sp. gr. 8–10; black with shades of green, brown, or grey, with resinous to greasy lustre; S. brownish-black, greyish to olive green.

In 1789 M. Klaproth, a German chemist, discovered a new element in uraninite, and he named it uranium, after the newly discovered planet Uranus. In 1896 the French physicist Henri Becquerel discovered that uranium emitted special radiation. This was the first element found with such properties, which he called radioactivity. Shortly afterwards Marie Curie discovered in uraninite the elements radium and polonium, which give more radiation than uranium. Both Becquerel and Marie Curie used uraninite from Jáchymov for their experiments. The Austro-Hungarian government gave permission, considering it a raw material of no particular value.

Uraninite generally forms compact vein filling, less often reniform aggregates. Crystal form is rare (in Norway and Sweden). It occurs in ore-veins, pegmatites, granites and coals. Uraninite is not a constant mineral. All too often many secondary minerals form from its alteration — mainly oxides, carbonates, phosphates and silicates, which have vividly bright colours of yellow, green, orange, etc.

The disintegration of radioactive elements became the basis of a new, extensive scientific field, which has given mankind an unsuspected major source of energy. The surface layers of uraninite veins, accessible to water and oxygen, are particularly rich in radioactive elements. Uraninite is distinct for its high density. The radioactivity of uraninite is extremely

196 Uraninite — Jáchymov (Czechoslovakia); 12 × 10.5 cm.

197 Uraninite — Shinkolobwe (Shaba, Zaïre); crystals 4 — 5 mm.

198  Uraninite — Shinkolobwe (Shaba, Zaïre); actual size of the illustrated section 9 × 6 cm.

strong. If placed with the ground surface on a photographic plate, a natural image of uraninite (a radiogram) will form through the emanation of disintegrated particles of the radioactive elements (16).

The largest uraninite deposits are in Canada, north of Lake Huron, and at Shinkolobwe (Shaba, Zaïre). This is where big, but imperfectly bounded crystals of uraninite occur, with faces up to 4 cm long. Crystals larger still are found in the locality of Wilberforce (Renfrew County), Ontario, Canada. They are either shaped like a cube, or an imperfect dodecahedron and are roughly 10 cm long. Uraninite occurs on all these deposits in the vicinity of granite massifs. In Sweden and other countries it is, however, extracted also from coal. Uraninite, which contains 75 to 96 per cent uranium, and some radium and polonium, is considered today the principal ore of these radioactive elements, on which the interests of the world are centred. It is a source of atomic energy both for peaceful and military purposes. As such, it has become the most important strategic raw material. There are such negligible traces of radium in uraninite (radium is used in medicine), that from a whole wagonload of uraninite less than 1 gramme of radium can be recovered.

199 Goethite — velvet variety — Příbram (Czechoslovakia); actual size of the illustrated section 16 × 12 cm.

**Goethite** (199) was often found in the cavities of limonite and taken for one of its varieties. In 1806 it was identified as a separate mineral, and named goethite in honour of the famous German poet, Goethe, who was also a collector of minerals. In colour and chemical composition it resembles limonite. However, in contrast to limonite, it forms small crystals, which are usually fine and needle-shaped. The rusty brown colour and velvety smoothness of crystals are typical of 'the velvet variety' found in Příbram in Bohemia. Goethite occurs with limonite through the weathering of pyrite, or may be formed from hot solutions (hydrothermal origin), or by dehydration of limonite. The most noted deposits are in Siegerland, West Germany, and in Cornwall, Great Britain. In the Slovakian Ore Mountains it is mined simultaneously with limonite. Economically it is of little value.

Hydrous iron oxide, $FeO(OH)$, orthorhombic. H. 5–5.5; Sp. gr. 4.3 (in aggregates it falls to 3.3); black-brown, red-brown, with adamantine to metallic lustre; S. brownish-yellow, orange-yellow.

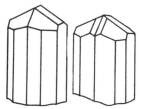

The striking appearance makes it a most desirable item for any mineral collector. Examples of the 'velvet variety' extracted from the surface layers of the Příbram mines which are now exhausted, can be seen in all major museums of the world. At one time this variety was called 'pribramite'. Wurtzite from Příbram was later also given the same name, so eventually the title was dropped for both the minerals.

183

200 Manganite
Ilfeld (German Democratic
Republic); 10 × 9 cm.

**Manganite** (200 — 201)

Hydrous manganese oxide
MnO(OH), monoclinic. H. 4;
Sp. gr. 4.33; steel-grey to black;
submetallic lustre; S. red-
brown to black.

was known in ancient times and was used in the decolorization of glass,
according to ancient historians. The yellowish colour of the crude vitreous
matter was aptly balanced by the gentle violet colour of manganese
compounds, which were added. At first manganese compounds were not
distinguished from similar iron ores, but in the second half of the 18th
century their difference was recognized. Manganite is usually found in
association with pyrolusite and psilomelane (manganese oxide), often in
deposits with admixtures of iron. Manganite commonly forms columnar
crystals grouped in druses or it develops crusts and delicately acicular
masses. It occurs in independent ore-veins with barite and calcite,
generally in magmatic rocks. Every famous collection in the world
contains druses of columnar crystals of manganite from Ilfeld in the
Harz Mountains. It is also found at St Just (Cornwall), England and
Negawnee (Michigan) and Douglas County (Colorado), USA. Manganite
is one of the most important manganese ores, and contains more than
60 per cent metal.

201 Manganite
Ilmenau (German
Democratic Republic);
aggregate 3 × 3 cm.

184

**Limonite** (202) is in fact natural rust. It is common in occurrence and originates mostly through the weathering of other iron ores, whose surfaces it covers with rusty stains. Limonite is usually powdery or earthy, but also compact, with reniform or stalactitic surface. The earthy and powdery varieties are called ochres or clays, the compact, lump varieties, with a resinous lustre, stilpnosiderites (from the Greek 'stilpnos' — glossy and 'sideros' — iron). Limonite is distinguishable from haematite, with which it often occurs, by the conspicuous colour of its streak. The most noted deposits are in Sweden and Finland, where it is called 'bog-iron ore', or 'lake ore', since it is found on the beds of some lakes. These ores are formed by the action of minute organisms. As the content of iron is relatively poor, it pays to recover limonite only where it is found in large quantity, or when it is associated with other iron ores.

Hydrous oxide of iron FeO(OH) +nH$_2$O, colloidal. H. changeable, most commonly 4—5.5; Sp. gr. 2.7—4.3; brown to yellow, with vitreous to dull lustre; S. brown.

202 Limonite — Rožňava (Czechoslovakia); 13 × 8 cm.

**Bauxite** (203) is the chief source of aluminium, which is rightly called the metal of the 20th century. Aluminium is one of the latest metals to be successfully extracted from an ore. Small amounts of aluminium were produced in 1886, but no one then thought that this could be of real significance, because the manufacture was excessively difficult. Much later though, aluminium was successfully produced by electrolysis. Bauxite commonly occurs in earthy granular or marbly (pisolitic) masses. It results from the decomposition of feldspathic rocks, weathered under tropical conditions. It is named after the locality where it was first found — the town of Les Baux, near Arles, in southern France. Bauxite is not composed solely of a mixture of aluminium oxides. It also contains iron oxides and an opal substance.

Sedimentary rock, whose chief components are aluminium hydroxides; white to brown shades.

There are large bauxite deposits in Yugoslavia, Hungary and USA (Arkansas, Alabama, Georgia), but the most important are in Jamaica and Guyana. The aluminium contents vary, but are usually between 25 and 30 per cent. Today aluminium is mainly used in the aircraft, automobile and shipping industries, and also in fuel elements for atomic reactors.

203 Bauxite — Les Baux (France); 8 × 6 cm.

# Chapter 5    CARBONATES

The salts of carbonic acid are called carbonates and they are a large and important group. Their origin in nature is varied and their occurrence is governed by the fact that they are a substance with very little resistance to acids and that they decompose at high temperatures under comparatively low pressure. They are divided into anhydrous and hydrous, or alkaline. The anhydrous carbonates fall almost entirely into two large groups, whose members mutually intermingle. The first is the rhombohedral calcite group (calcite, magnesite, siderite, rhodochrosite, smithsonite) and the second is the orthorhombic aragonite group (aragonite, cerussite, and others). Double salts (e. g. dolomite), which crystallize also rhombohedrally, but slightly differently, are closely allied to the calcite group. The alkaline hydrous carbonates contain chiefly carbonates of bivalent metals, such as copper, lead and zinc. Natural occurrences of nitrates and borates are commonly included with carbonates.

**Dolomite** (204) sometimes forms whole mountain complexes. It is one of the carbonates which resemble other carbonates so closely in various characteristics that originally they were not distinguishable. All minerals of this appearance were therefore considered for many years to be calcites, and the rock limestone. This was changed in 1791, when the French geologist D. Dolomieu pointed out various different properties of a rock which occurred mainly in southern Tirol. The Swiss mineralogist H. B. de Saussure studied the rock in closer detail (1740 — 1799) and named it dolomite in honour of Dolomieu. With its crystal shape, chemical composition and colour, dolomite resembles calcite and magnesite. Whole mountain ranges or large areas are sometimes composed of it, e.g. magnesium limestone in England. Dolomite rocks originally occurred by deposition of shells of small sea animals. Their origin is very similar to that of limestone rocks. Crystallized dolomite is most commonly found in ore-veins, for example at Solbad Hall (Tirol, Austria), in the Binnen Tal (Switzerland), in Banská Štiavnica (Slovakia) and in Vermont (USA). Dolomite is used in the manufacture of special cements and for making refractory linings.

Calcium magnesium carbonate, $CaMg(CO_3)_2$, rhombohedral. H. 3.5–4; Sp. gr. 2.8–2.9; grey, white, reddish, less commonly dark, with vitreous to pearly lustre; S. white.

204 Dolomite
Banská Štiavnica
(Czechoslovakia); 13 × 9 cm.

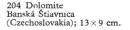

**Azurite** (205) occurs in association with malachite from the weathering of any copper ores, and they are both responsible for the typical surface colouring. The crystals of azurite are of an exceptional, deeply blue colour and its granular, earthy or pulverulent covers are the colour of forget-me-nots. The richest deposits of azurite are at Tsumeb in Namibia, at Bisbee and Morenci in Arizona (USA), and Burra near Adelaide in Australia. In Europe it is recovered at Chessy near Lyon, France, and at Băita in Romania. Azurite, which contains 55 per cent copper, is rarely extracted as an ore, because it hardly ever occurs pure. When ground, it can be used as a blue pigment and in the production of blue vitriol. It is too soft to be much used as an ornamental stone. But when it is cut and polished, it has a strong lustre and beautiful shades. Vividly colourful specimens of azurite with malachite are most popular among collectors.

Basic carbonate of copper, $Cu(OHCO_3)_2$, monoclinic. H. 3.5—4; Sp. gr. 3.7—3.9; blue, vitreous lustre; S. light blue.

205 Azurite — Tsumeb (Namibia); larger crystal 10 cm.

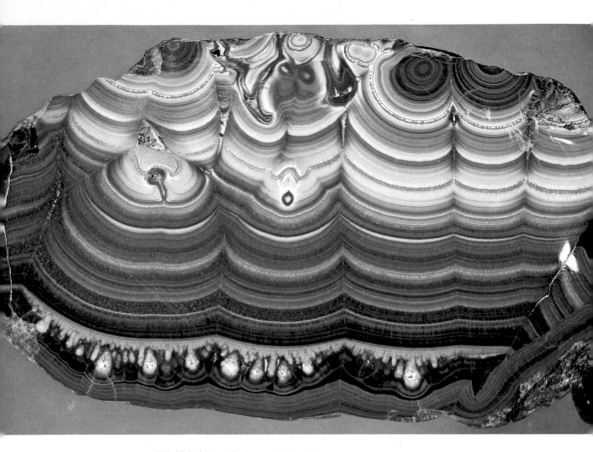

206 Malachite — Shaba (Zaïre); 13 × 7 cm.

**Malachite** (206) is not just a modern, ornamental stone. The Greeks and the Romans used it in the manufacture of amulets, which, according to their beliefs, were to protect people, especially children, from various misfortunes. Malachite was also found by ancient miners in copper mines, in the form of vivid green coatings, which at first they called 'rock green'; it was later given its present name. Malachite has had other names, such as 'the velvet ore', or 'the satin ore'.

Basic carbonate of copper, $Cu_2(OH)_2CO_3$, monoclinic. H. 4; Sp. gr. 4.0; green to dark green; vitreous adamantine to silky lustre; S. green.

Malachite is the most common product of the weathering of copper and copper ores. It is abundantly present in the upper layers of all the deposits of these ores. Apart from the surface covers, malachite commonly forms fine, fibrous layers with a reniform surface, or powdery concretions. It occurs exceptionally in cavities in the form of needle-shaped crystals. The best known deposits of this mineral are in extensive regions of the Urals, especially **Gumeshevsk, Karpinsk** and **Nizhniy Tagil.** The biggest malachite boulder found in Gumeshevsk weighed 60 tonnes. Other large occurrences are in **Zaïre,** and it is also found in the United States in **Bisbee, Gloke, Clifton (Arizona), Park City (Utah), Good Springs (Nevada)** and in **Cornwall in England, Tsumeb in Namibia** and **Burra near Adelaide (Australia).** The most recent deposit of malachite to be exploited for ornamental purposes is at **Eilat in Israel.** It is possible that this newly discovered deposit is the forgotten mine of the ancient Greeks and Romans. Malachite is a semiprecious stone (3), but also a valuable ore (it contains approximately 57 per cent copper).

189

207 Trona (sodium) — southern California (USA); grains 4 mm.

**Trona (sodium)** (207) forms granular, needle-shaped crusts, earthy coatings and concretions in the sediments of so-called 'bitter sodium lakes'. Largest deposits of natural sodium are in the USA (California, Nevada), and in the USSR (Armenia). Sodium is used in chemical industry.

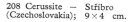

Hydrous sodium carbonate, $Na_2CO_3 . 10H_2O$, monoclinic. H. 1–1.5; Sp. gr. 1.45; colourless, white to grey, silky lustre; S. white.

**Cerussite** (208) was first found when mining surface silver and copper ores. The upper parts of ore-veins were the location of the magnificent specimens of crystallized cerussite. The name comes from the Latin 'cerussa' — 'white paint'.

Lead carbonate, $PbCO_3$, orthorhombic – pseudohexagonal. H. 3–3.5; Sp. gr. 6.4–6.6; greyish, white, greyish-black, with almost adamantine lustre, sometimes somewhat greasy; S. white.

Cerussite always originates as a product of the weathering of the mineral galena, especially in parts of ore-veins which are easily accessible to surface waters and carbon dioxide. Water frequently acts as a solvent to many minerals and galena does not escape this process. Cerussite crystals have various forms, and they are often intertwined in various arrangements. They have an exceptionally strong lustre. The rich druses of cerussite crystals from German deposits adorn collections all over the world. There are also abundant deposits in Nerchinsk in the USSR and Colorado in the USA. Cerussite contains approximately 73 per cent lead, but it is valuable as a lead ore only in very few localities, where it occurs in extensive quantities.

208 Cerussite — Stříbro (Czechoslovakia); 9 × 4 cm.

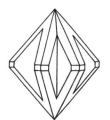

190

209 Aragonite — iron flower Eisenerz (Austria); 20 × 15 cm.

**Aragonite** (209—213) was first discovered in the 15th century, when on Mount Erzberg in Austria a gallery was driven across a rich deposit of siderite ores. This gallery was intended to simplify the search for silver and gold veins. The new mineral looked at first sight like a snowy white, or faintly blue plant turned to stone, and was appropriately called 'the iron flower' (209). Basically these are mutually intertwined stalk-like aggregates.

Calcium carbonate, $CaCO_3$, orthorhombic. H. 3.5–4; Sp. gr. 2.95; commonly white, grey, with reddish to black shades, with vitreous lustre; S. white.

Linnaeus discovered that this 'iron flower' is chemically identical to calcite. Later experiments proved, however, that it has a number of properties which identified it with what were supposed to be calcite crystals in Aragon (Spain). From then on the mineral was named aragonite.

Aragonite is distinguishable from calcite in that it has a very poor cleavage and does not occur in rhombohedral form. Its columns are

210 Aragonite — Příbram (Czechoslovakia); 11 × 11 cm.

211 Aragonite — peastone — Karlovy Vary (Czechoslovakia); 17 × 12 cm.

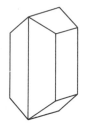

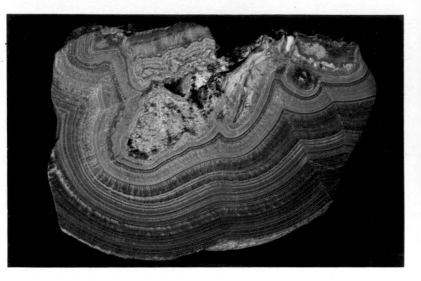

212 Aragonite — sprudelstein — Karlovy Vary (Czechoslovakia); 7 × 5 cm.

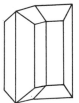

usually intergrown into large six-sided forms. The crystals are often acicular to fibrous. Like calcite, aragonite also occurs under the most varied conditions. Calcite, however, crystallizes from relatively cool solutions whereas aragonite crystallizes at higher temperatures. This affects the inner structure, which differs from calcite and conditions its diverse crystal form and diverse physical properties. Deposits from hot springs cause the origin of aragonite's fine lumpy variety — **peastone** (211) which occurs in Karlovy Vary in Bohemia, and resembles pea grain. Some very fine crystallized examples come from Tirol. Crystallized aragonite has no practical importance but is much sought after by collectors. The **sprudelstein** (212) found in Karlovy Vary is composed of rich brown bands of various shades and has been used for a long time for the manufacture of ornamental objects.

213 Aragonite — Schwaz (Tirol, Austria); 8 × 2 cm.

**Calcite** (214—221) aroused no interest until the second half of the 17th century, when a very strange discovery was made near Eskifjördur on the eastern shore of

Calcium carbonate, $CaCO_3$, rhombohedral. H. 3; Sp. gr. 2.6—2.8; commonly colourless, white, brownish, yellow, blue, and of other shades, but rarely dark; vitreous lustre, pearly on cleavage planes; S. white.

Iceland. While extracting basalt, stone workers hit a cavity measuring $6 \times 3$ m, which was filled with a multitude of grey-white spindly calcite crystals. Some were completely clear, others only partially. The recovered calcite excited attention by its shape and colouring and, as a so-called 'silverstone', was exported into Europe.

Erasmus Bartholin, a Danish surgeon, examined this mineral most carefully and published his findings in 1669. He wrote of his discovery that there is pronounced double refraction when light passes through this mineral. If, for instance, we cut out a rhombohedron from a clear calcite crystal, and look through it at some writing on paper, we will see the writing double. The discovery of double refraction proved valuable in optical research.

Rocks formed by calcite limestones were, of course, known a long time ago. Granular limestone, or **marble**, was for instance the main material used by ancient Greek sculptors. It was extracted mainly from Paros and the mountain of Pentelikon near Athens, and just as Greece used to be, Italy was and is still renowned for its marble. During the 15th century mining for marble began in earnest on the slopes of the Apennines, especially near Carrara. In the Middle Ages marble was used mostly to decorate churches and for sculpture. The so-called 'marble of the ruins' (221) from Florence was very popular at that time.

214 Calcite — Eskifjördur (Iceland); crystal 6 cm.

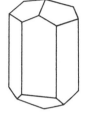

215 Calcite — Dürkheim (Federal Republic of Germany); 8 × 5 cm.

Apart from quartz, calcite is the most common mineral and occurs in an almost endless variety of forms. Between crystals bounded by perfect faces and extensive calcareous beds formed of minute microscopic grains (218) of different geological ages, there exist hundreds of forms and combinations. Calcite usually crystallizes in rhombohedral form of perfect cleavage, which as a rule are grouped in druses. The massive variety, fine to coarse grain, or compact rocks, are far more useful. These rocks have various names. **Travertines** (219), for instance, are porous limestones deposited from calcite waters, and marbles are crystalline limestones, which are often beautifully coloured. **Marl** and arenaceous marl are compact limestones with an admixture of clayey and sandy

216 Calcite — Baia de Aries (Romania); 10.5 × 8 cm.

217 Calcite — Egremont (Cumbria, England); 13 × 9 cm.

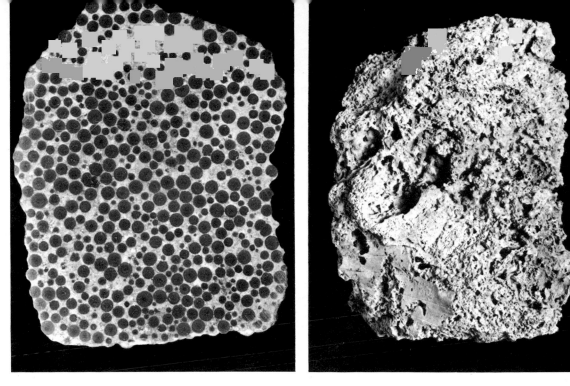

218 Oolitic limestone —
Mannsfeld (German Demo-
cratic Republic); 12 × 9 cm.

219 Travertine — Trenčianské Teplice (Czechoslovakia); 12 × 8 cm.

220 Calcite — nail-head spar
— Příbram (Czechoslovakia);
7 × 7 cm.

particles. All these outwardly distinctly dissimilar varieties of one particular mineral resulted from the diversity of conditions during their development. The crystallized calcite forms in ore-veins, in cavities of calcareous rocks and in cavities of various igneous rocks. There it forms chiefly from the sedimentation of hot solutions. It is therefore possible to find crystallized limestone almost everywhere. The majority of calcareous rocks occurred and still occur through sedimentation on the ocean bed. This is because rivers carry into the seas and oceans a mass of minute particles of calcium carbonate from the weathered rocks of the earth's surface. This is necessary for the formation of some animal skeletons and shells. As the animals die, the remains of skeletons and shells keep amassing.

The most important deposits of calcite crystals are (apart from Iceland) in Joplin County (Missouri) and Toos County (New Mexico), USA. The European deposits are in England at Ely, Weardale, Matlock and Eyam (Derbyshire), Egremont (Lake District) and Liskeard (Cornwall), near Fontainebleau in France, in Bavaria and Romania. The so-called 'nail-head spar', a characteristic mineral of Příbram and of Stříbro in Bohemia, is much sought after by collectors. It is formed by an interesting grouping of parallel twins of low rhombohedral crystals, where the upper parts resemble nail-heads. The use of marble has already been mentioned. But other calcite varieties also have a technical importance and are an essential raw material. The clear crystal varieties are widely used in optical instruments, for which they are highly suitable because they have double refraction.

221 'Marble of the ruins' — Florence (Italy); 14 × 9 cm.

222 Rhodochrosite — Capnic (Romania); aggregates 12 mm.

**Rhodochrosite** (222) (from the Greek 'rodon' — a rose, and 'hrós' — colour), or **dialogite** (from the Greek 'dialoge' — choice) is most commonly found in ore-veins and sedimentary deposits of pyrite and manganese ores. Here it commonly forms small veins with pyrite and psilomelane. Less commonly it occurs in cracks of igneous rocks, such as spilites. The crystal form is rare; usually rhodochrosite is massive, coarse or fine-grained to compact. Sometimes it forms globular, reniform, botryoidal aggregates and concretions.

Manganese carbonate, $MnCO_3$, rhombohedral. H. 4; Sp. gr. 3.3–3.6; commonly rose to scarlet, almost white, dirty grey, brownish, greenish, rarely colourless, with vitreous lustre; S. white.

Some magnificent crystals of rhodochrosite come from the Harz Mountains, Freiberg in Saxony, Romania, the western states of the USA, e.g. Alicante (Colorado), and Siberia. Commercially useful deposits occur only in the Pyrenees and near Huelva, southern Spain. Here rhodochrosite is recovered as a manganese ore, containing up to 42.8 per cent manganese. Manganese is used in making a particularly hard steel and numerous manganese compounds are of practical importance. The crystals from the Argentine deposits are used as precious stones.

197

223 Magnesite — Graz (Austria); 11 × 8 cm.

# Magnesite

**Magnesite** (223) was so named because it contains magnesium, but for a long time it was not given any particular attention. Up to 50 years ago magnesium metal was used solely as an ingredient of fireworks or as a source of light during inferior lighting conditions in photography. With the development of modern industries its uses grew and keep on growing.

Magnesium carbonate, $MgCO_3$, rhombohedral. H. 4—4.5; Sp. gr. 3.0; white, yellowish or greyish, with vitreous lustre; S. white.

Like the carbonates, magnesite also occurs in many varied forms, either crystalline or compact. When crystalline it is fine to coarse-grained, outwardly closely resembling marble. Granular magnesites occur through alterations of limestone and dolomite, though on rare occasions also by direct deposition. The compact varieties form through decomposition of serpentines. The latter often contain admixtures of opal and are consequently hard.

The chief European magnesite localities are in Austria (Veitsch, Leoben and Mürzzuschlag, Styria), in the southern Slovakian parts of Czechoslovakia, in the USSR (Satka in the southern Urals), in Greece and Italy (Tuscany). It also occurs in the northwest of the USA (California, Nevada, Washington), and in New South Wales, Australia. Magnesite is extensively and widely used. It is an important raw material in the building industry, and also important for its fire-resisting properties. Heat-resisting bricks are made from it and it is used for the manufacture of furnace linings. It is also used in the production of porcelain, ceramics, and also in the manufacture of quick-setting cement.

198

224 Siderite — Příbram (Czechoslovakia); 8 × 7 cm.

**Siderite** (224) is an important iron ore, which was being recovered from Mount Erzberg in the Eisenerz Alps in Styria, Austria, as long ago as the beginning of the 8th century. This mountain, where mining still goes on today, is completely composed of siderite. Its origin is most interesting. Originally it was formed of limestone, but hot solutions rising from the earth's interior simultaneously carried away and deposited matter and gradually replaced the limestone with the less soluble siderite (hydrothermal metasomatism). During this process the so-called 'iron flower' (209) formed from the dissolved calcium carbonate.

Iron carbonate $FeCO_3$, rhombohedral. H. 4—4.5; Sp. gr. 3.7 to 3.9; commonly brownish-yellow, dark-brown to black, sometimes with bright metallic shadows; vitreous, at times even pearly lustre. S. white.

Siderite frequently forms veinstones, or occurs in association with other ores, copper for instance. In such deposits it is massive or crystallized. The best examples of siderite's rhombohedral crystals, clustered in neat rose-shaped forms and druses, frequently come from ore-veins. Such occurrences, however, are of low economical value. The occurrences of clay siderite (pelosiderite), which forms balls and nodules in some coal basins, are most interesting. They are often nearly half a metre long (northern and southwest England; Halle in East Germany). Though it contains only a relatively small proportion of iron (48 per cent), siderite is an important iron ore, for in comparison to other iron ores, it is usually pure and is easy to work.

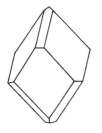

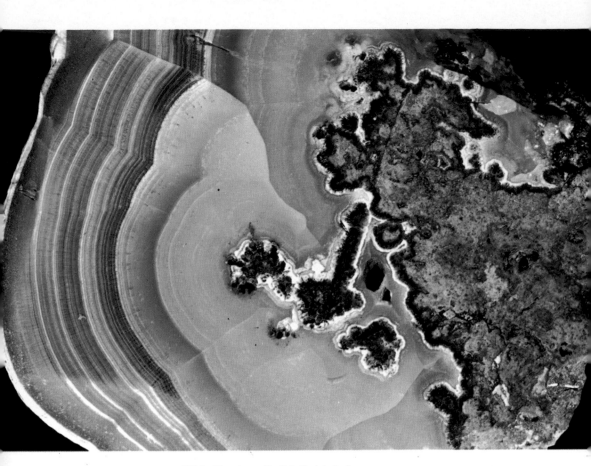

225 Smithsonite — Sardinia (Italy); 9 × 6 cm.

**Smithsonite** (225) was mined very early as a zinc ore, though zinc was not then used on its own.

Zinc carbonate, $ZnCO_3$, rhombohedral. H. 5; Sp. gr. 4.3–4.5; commonly grey-white, colourless, whitish, yellowish, with green or blue shades, brown; strong vitreous lustre, even slightly pearly; S. white.

There are many beautiful brass objects extant from these days which confirm that brass (an alloy of copper and zinc) must have been produced and in large quantities. Many minerals, which when mixed with copper produced the popular brass, were then well known. This applies especially to so-called 'calamine', which occurs in the upper layers of zinc and lead ore-veins. Not until the second half of the 18th century was it discovered that calamine was two different minerals, zinc carbonate (smithsonite) and zinc silicate (hemimorphite).

Smithsonite was named after J. Smithson (1765–1829), a British chemist and mineralogist. It forms reniform, botryoidal and rounded aggregates, or earthy crusts and fibrous clusters. As it is a product of the surface weathering of sphalerite, it occurs mainly in the upper layers of its beds. The main deposits are in Polish Silesia, Sardinia and Nerchinsk in Siberia and also Matlock in England, Tintic (Utah), Mineral Point (Wisconsin) and Kelly (New Mexico). The most beautiful translucent compact smithsonites in existence come from the copper ores of Tsumeb in Namibia. Smithsonite is an important ore of zinc, easily reducible, and containing up to 52 per cent zinc.

# Chapter 6      BORATES

Borates are the salts of boric acid. They are of non-metallic appearance, mostly soft. They originate in places with high concentration of the element boron; they occur in the earth's crust, in sea water and its sediments, in magmatic exhalations of varied composition. Borates are found in granite pegmatites (anhydrous borates), in deposits of contact and metamorphosed rocks, in salt deposits, in sediments of geysers and lakes as well as in the products of fumaroles (exhalations of hot vapours and gases on active volcanoes). Borates are divided into hydrous and anhydrous. They are not commonly distributed, though, locally, they form technically important deposits. Borax and kernite are the most important.

**Borax** forms through sedimentation in some lakes, for example in the Clear Lake in California, in Nevada, near Arequipo in Peru, in western Tibet, in Iran and in several places in central Africa. It is also found in volcanic exhalations in Tuscany (Italy) and on some volcanoes in the Kerch strait (USSR). In favourable conditions it forms perfectly cleavable columnar crystals and is extracted through water evaporization from mineral lakes, especially in California and Tibet.

Hydrous sodium tetraborate, $Na_2B_4O_7 \cdot 10H_2O$, monoclinic. H. 2—2.5; Sp. gr. 1.7; usually white with silky lustre; S. white.

**Kernite**, mixed with clay, forms vast deposits in California, similarly to **colemanite** (226) which often forms large, multifaced crystals. Its main deposits are Death Valley in California (USA) and Panderma in Turkey. Boron compounds are used mainly in medicine and chemical industry, in the manufacture of soap and glass, in melting porcelain mass and in the manufacture of modern washing and leaching preparations. Boron alloys are used for regulation bars in nuclear reactors and for the prevention against neutron radiation.

Hydrous sodium tetraborate, $Na_2B_4O_7 \cdot 4H_2O$, monoclinic. H. 2.5; Sp. gr. 1.9; usually colourless with silky lustre; S. white.

Colemanite: Hydrous sodium borate, $Ca_2B_6O_{11} \cdot 5H_2O$, monoclinic. H. 3.5; Sp. gr. 2.4; usually colourless, silky to adamantine lustre; S. white.

226 Colemanite — Panderma (Turkey); crystal up to 2 cm.

# Chapter 7     SULPHATES AND SIMILAR COMPOUNDS

Sulphates are the salts of sulphuric acid. They are soft and of non-metallic appearance. They are usually the result of deposits from sea water, or of volcanic eruptions. Sometimes they result from the action of sulphuric acid which was formed during the oxidation of sulphides, mainly pyrite and marcasite. Sulphates are divided into anhydrous (such as anhydrite and barite) and hydrous (such as gypsum and melanterite). Several interesting groups of minerals belong to the sulphates, for instance vitriols (binary hydrous sulphates of monovalent and trivalent metals) and some binary hydrous sulphates of bivalent and trivalent metals.

**Celestite** (227), sometimes called **celestine**, was a familiar mineral to Bengalese priests, who used

Strontium sulphate, $SrSO_4$, orthorhombic. H. 3–3.5; Sp. gr. 3.9–4; commonly blue, colourless, yellowish, reddish, with vitreous to pearly lustre, when broken somewhat greasy; S. white.

it to colour flames. With its help they prepared vividly crimson Bengalese fires, whose mysterious beauty filled the simple believers with awe and with horror. Not till much later did chemists discover that this flame colouring was caused by the element strontium, a component of celestite, while its relative, barium, tints the flame yellow-green. Chemists obtained volatile salts of both these metals from certain minerals, and mixed them with potassium chloride, charcoal and sulphur, thus producing the very first firework rockets.

The name celestite comes from the Latin 'coelestus' — heavenly, for its sky-blue colour. It usually occurs in tabular crystals, or columnar crystals that resemble barite crystals in shape. It is commonly found in sedimentary rocks, in hydrothermal veins and in cavities of igneous rocks. In Sicily it occurs in crystal form. Industrially valuable deposits are at Yate, near Bristol in England, in Egypt, and at Lake Erie, Death Valley and Nashville, USA. Celestite is used in the chemical and food industries and for surgical and pyrotechnical purposes.

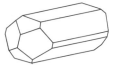

227 Celestite — Špania Dolina (Czechoslovakia); crystals 3 cm.

**Barite** (228—231) was familiar to the very first miners, for it occurs abundantly in ore-veins. For a long time it was considered to be worthless gangue, though it had stirred the interest of alchemists in the Middle Ages. Vincenzo Cascatiolo, an Italian alchemist and shoemaker in Bologna was heating barite with other minerals in the course of his experiments in 1630. He discovered that when heated, it phosphoresces in the darkness. This was the discovery of phosphorescence, an occurrence well known today, but one that in those days caused great excitement. As Cascatiolo studied the round clusters of barite from gypsum marl in the vicinity of Bologna, the stone he discovered became known among his contemporary natural scientists as 'the shining ball of Bolonga'.

Barium sulphate, $BaSO_4$, orthorhombic. H. 3–3.5; Sp. gr. 4.48; white and variably coloured, commonly shaded green, yellow, red, blue or brown; vitreous lustre, pearly on some planes; S. white.

Much later, the nodules were examined in more detail, and then it was discovered that the 'shining balls' were the same mineral as the well known, heavy stone from the ore-veins. Barite received its name from the Greek 'barys' — heavy, for it was conspicuously and unusually heavy. The high density of barite nears the density of some of the iron ores, magnetite or haematite for instance.

Barite occurs abundantly in cavities of ore-veins, in association with sulphide ores, especially lead, zinc and silver. It was formed by deposition from hot solutions rising from the interior of the earth at higher temperatures. Often it occurs in veins, or in sedimentary deposits. Apart from its remarkable weight it is also distinctive by its often perfect crystals. They usually have the form of thick, orthorhombic plates and columns, and are frequently exceptionally large. Some of the columnar crystallized varieties, which differ to a certain extent from the normal

228  Barite — Cleveland (Oklahoma, USA); 4.5 cm.

229  Barite - Dědova Hora (Czechoslovakia); crystals 8—16 mm.

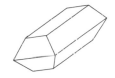

development of crystals, are classified as wolnyne. Compact barite forms various aggregates, mostly lamellar and of good cleavage.

The occurrence of separate barite veins, commonly containing an admixture of fluorite, happens usually in the closest association with granite massifs. The particular occurrences of compact barite found in such veins are the most important ones for practical use. The deposits in sedimentary rocks, where it often forms tubers, are also valuable. The colour is usually very pale. When pure, it is colourless to white, but commonly it is tinted by various admixtures.

The chief world producers of barite are the USA, West Germany (Meggen deposit), England (Cumbria, Cornwall, Surrey, Derbyshire),

230 Barite — Freiberg (German Democratic Republic); 13 × 9 cm.

and the USSR. Compact barite is commercially useful. But from the mineralogical and collectors' point of view, the occurrence of barite crystals at Egremont (Cumbria, England), Baia Sprie (Romania), Freiberg (Saxony, East Germany) and in Czechoslovakia, are much more interesting.

The present-day chemical industry makes the fullest use of barite especially in the manufacture of paints. In particular, the compound barium sulphide, forms the base of white paint that is completely non-poisonous and resistant to chemical effects. When ground, barite is used as a filling for paper and other substances, to give them weight. The ceramic industry uses barite in the manufacture of glazes and enamels. It is also used in the glass industry and in the production of barite cement and heavy concrete for protection against radioactive rays. Barite as a raw material is also much sought after for the production of colouring compounds needed in medicine. The use of volatile barium salts for tinting flames (Bengalese fires), for rockets and fireworks has already been mentioned. Collectors of minerals were quick to appreciate the beauty of the rich barite druses, which are not just lovely and interesting, but are frequently formed by exceptionally large crystals of the most diverse colours. The specimens of barite crystals from many British localities, particularly those found in Cumbria and Tavistock, are the ones most highly prized. From these noted deposits come large tabular crystals, pointed at each end, with the prevalent colours of greyish-yellow, blue-grey to honey-yellow. These deposits are also known for their slender columnar crystals which are equally beautifully coloured.

231 Barite — Baia Sprie (Romania); 12 × 7 cm.

232 Gypsum — selenite — Sverdlovsk (Urals, USSR); 8 × 5.6 cm.

**Gypsum** (232—235) has been known since antiquity. The sculptor Lysippus from Sicyon was the first to use plaster manufactured from gypsum. According to Pliny the Elder he made the very first plaster cast of a human face.

Hydrated calcium sulphate, $CaSO_4 . 2H_2O$, monoclinic. H. 1.5–2; Sp. gr. 2.3–2.4; commonly colourless, white, yellowish, reddish to blood-red, grey to black, with vitreous lustre, pearly on cleavage planes, or fibrous and silky; S. white.

It is not generally known that gypsum was probably the first mineral to be studied under the microscope in 1695 by Anton van Leeuwenhoek, a Dutch amateur natural scientist, the founder of microscopy.

Gypsum occurs in tabular to columnar crystals resembling mica ('Maria glass'). In its pure crystallized form it is sheer and colourless. Its massive variety is white, or often tainted yellow or brown, according to the admixture present. When finely granular it is called **alabaster**; when clear white and transparent, it is called **selenite** (232—233); when fibrous, it is **satin spar**. Gypsum is distinguishable from the others because it can be scratched with the finger nail.

The mineral is formed by deposition from sea water, often in association with anhydrite, which often alters to gypsum through water absorption. Interesting occurrences of gypsum originate also through the weathering of pyrites, for instance in ore-veins and in coal seams.

The most noted gypsum deposits are near Volterra, Tuscany, in Spain and Egypt, where it occurs abundantly. Other large deposits are in the USA, USSR, France, Britain (Matlock, Swanage, Oxford, Isle of Sheppey), and Wieliczka in Poland, also in Czechoslovakia (Opava) and in a number of other places. In many localities in the Sahara, perfectly developed agglomerates of gypsum crystals occur lying loosely in the sand. This gypsum variety sometimes contains more than 50 per cent

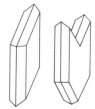

206

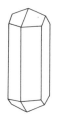

fine grains of sand, which it drew into its brownish, greyish, or yellowish lenticular crystals during rapid crystallization. These interesting formations are commonly flat, quite often fairly large, grouped into shapes resembling roses in bloom, or other flowers and leaves. This is why they are called 'desert roses'. Such formations originated through evaporation of water from salt lakes, or through the decomposition of pyrite.

Gypsum's main use is in the production of plaster. When heated to 300—400°C it yields plaster of Paris, which sets hard after being mixed with water. It is used in the building industry, in medicine, and for making casts. When gypsum is heated to more than 400°C, it does not absorb water. It is then used in paints and cement. Alabaster is suitable for sculpture; selenite, highly popular for its unusual lustre, is a semi-precious gem. Less frequently gypsum is used for the production of sulphuric acid and sulphur.

**Anhydrite** (236) is similar to gypsum for which it had been originally mistaken in appearance and

Calcium sulphate, CaSO₄, orthorhombic. H. 3.5; Sp. gr. 2.9–3.0; commonly white, colourless, bluish, greyish, reddish, on cleavable planes pearly, otherwise vitreous to greasy lustre; S. white.

chemical composition. The name came from Greek ('an' — without, 'hydor' — water), for in contrast to gypsum, anhydrite does not contain water. It is usually compact, and only exceptionally cleavable. Anhydrite deposits occurred directly from sea water, like the deposits of rock salt and gypsum. This is why anhydrite is often associated with gypsum, and sometimes alters to gypsum through water absorption. Like gypsum but not so rarely, it also originates through the weathering of pyrites.

The most important deposits are at Wieliczka in Poland, Salzburg in Austria, Berchtesgaden and Lüneburg in West Germany, Stassfurt in East Germany, Vulpino near Bergamo in Italy, Spišská Nová Ves in Slovakia, the states of New York, Kansas and New Jersey, USA, and Nova Scotia and New Brunswick, Canada.

233 Gypsum – cut selenite – Sverdlovsk (Urals, USSR); 9 × 3 cm.

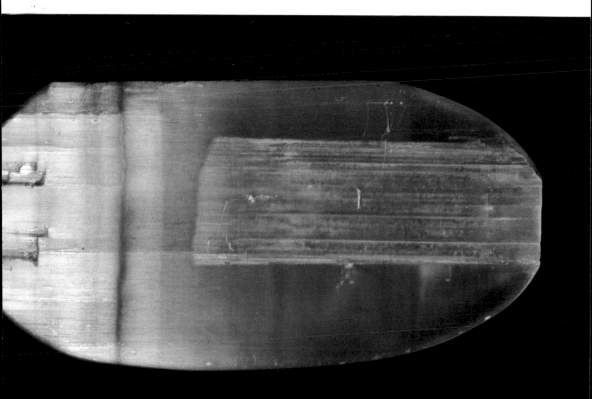

234 Gypsum — Bochnia (Poland); crystals 5 cm.

235 Gypsum — Čermníky (Czechoslovakia); 14 × 10 cm.

236 Anhydrite — Wieliczka (Poland); 12 × 8 cm.

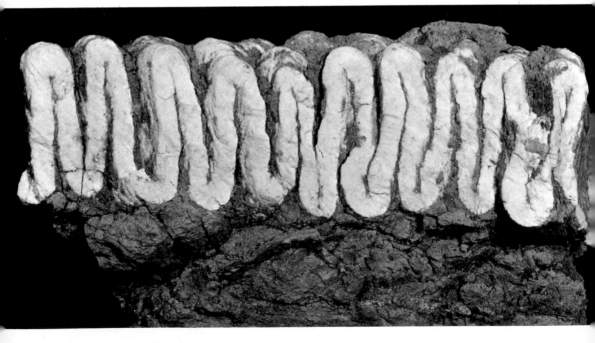

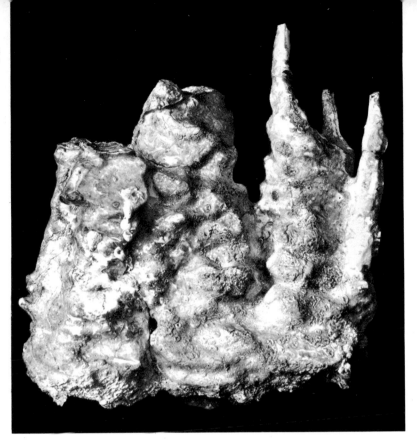

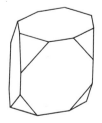

237 Melanterite — Smolník
(Czechoslovakia); 16 × 16 cm.

**Melanterite** (237), the natural green vitriol (copperas), is yet another mineral which originates through the weathering of pyrites. During the decomposition of pyrite, sulphuric acid is freed, and decomposes soluble minerals in the vicinity, and then in association with them forms various sulphates. Such minerals are often conspicuous in colour. The name melanterite is derived from the Latin term 'melanteria', which was used by Pliny. Melanterite usually forms stalactitic crystals or powdery and fibrous aggregates and concretions. This interesting mineral has to be taken care of in a special way, when part of a collection. It is very susceptible to weathering, losing its water and oxidizing very easily. It must therefore not be exposed to excessive heat or dryness, but also must not be subjected to damp. Otherwise the specimens, naturally beautifully green and translucent, will start to disintegrate and lose colour, turning yellowish.

Hydrated iron sulphate, $FeSO_4 . 7H_2O$, monoclinic. H. 2; Sp. gr. 1.89–1.90; pale green to white, grows yellow with oxidization; vitreous lustre; S. white.

The best melanterite deposits are in West Germany (Rammelsberg near Goslar), in Sweden (Falun), in Yugoslavia (Idria), in Slovakia (Banská Štiavnica) and in Utah (Bingham,) Tennessee (Ducktown), and California (Alma). As melanterite does not occur abundantly, it is of little practical value, though of course green vitriol is important. It is used for instance for the manufacture of ferric paints. For this purpose larger amounts are needed, so it is obtained mainly from pyrite.

**Mirabilite** was given the name from the Latin 'mirabilis' — admirable, for its healing properties. It originates in the form of crusts and coatings of strikingly salty to bitter savour from sediments of bitter lakes or mineral waters. The main deposits are at Laramie, Wyoming (USA), Canada, Chile and the USSR (Kara Bugaz Bay in the Caspian Sea). Mirabilite is used in the glass industry and for the manufacture of sodium.

Mirabilite:
Hydrous sodium sulphate, $Na_2SO_4 . 10H_2O$, monoclinic. H. 1.5–2; Sp. gr. 1.4–1.5; white with silky lustre; S. white.

209

238 Halotrichite — Dubník (Czechoslovakia); 8 × 5.5 cm.

**Halotrichite** (238) was

Hydrated aluminium ferrous sulphate, $FeAl_2(SO_4)_4 . 22H_2O$, monoclinic. H. 2.5; Sp. gr. 1.8 to 1.9; colourless, with silky lustre; S. white.

given the name from the Greek 'als' — salt, and 'trichos' — hair. It is yet another water-soluble sulphate which often occurs during decomposition of pyrite. It is one of the most common natural alums. It is often associated with its relative, pickeringite, water-soluble magnesium aluminium sulphate. Halotrichite crystallizes in capillary, fibrous, acicular crystals, which usually form in slates. The main deposits: Mörsfeld (Germany), Červenica-Dubník in Slovakia, Gila River in New Mexico (USA), Copiapó in Chile, and Rezaiyeh in Iran. Halotrichite is not as yet used commercially.

Another interesting mineral from the alum group is **chermikite** (the name comes from the deposit at Čermníky, Bohemia), hydrous potassium amonium sulphate. It forms layers in clay and coal pit-heaps, generally with numerous other sulphates. The tabular layers are composed of fibres which are oriented perpendicularly towards their surface. The crystallized form is rare.

**Bilinite** is yet another natural alum to be named after the locality of its discovery (Bílina, northern Bohemia). This monoclinic hydrous ferric-ferrous sulphate is generally yellowish and forms finely radial fibrous aggregates in the brown coal deposits near Bílina.

**Cyanotrichite** or **lettsomite** (239) occurs similarly to the previous mineral and is a member of the same group. But it is much rarer. It forms velvety aggregates of extremely fine fibres and radial aggregates. The main deposits: Cap la Garonne in the Var (France), Romania and Arizona (USA). Because of its beautiful colouring, lettsomite is very popular with collectors.

Complex hydrated basic sulphate of aluminium and copper, $Cu_4Al_2(OH)_{12}SO_4 . 2H_2O$, orthorhombic; H. 1; Sp. gr. 2.7–2.95; azure blue to ultramarine, with silky lustre; S. light blue.

239 Cyanotrichite — Moldova (Romania); 6 × 4 cm.

**Zippeite** (240) caught the attention of a Bohemian metallurgist, Adolf Patera in the 1850s by its striking colour. This was during the period when the fame of the Jáchymov silver mines was dying out. On top of the large, forgotten pit-heaps, powdery minerals began to appear, vividly coloured, originating from the decomposition of the pitchblende. These colourful layers, in which zippeite was in the majority, gave Patera the idea of utilizing the uranium minerals in the manufacture of paints. Mining began in 1859 not only for the products which resulted from uraninite's decomposition, but for uraninite itself. A variety of yellow paints were manufactured because among all the secondary minerals on the pit-heap there was a prevalence of the yellow powdery coatings.

Complex hydrated sulphate of uranium, $U_2SO_{10} + 3-6H_2O$, monoclinic. H. 3; Sp. gr. 2.5; greyish-yellow to lemon yellow; S. yellow.

Zippeite is one of the so-called 'uranium ochres'. It occurs in association with uranopilite, a monoclinic, complex water-soluble alkaline sulphate of uranium, from which it hardly differs in appearance, and with other secondary uranium minerals in the weathered veins of uranium. It was named after the Prague mineralogist, F. X. M. Zippe (1791—1863). Apart from Jáchymov in Bohemia it occurs mainly near Wölsendorf in Bavaria (West Germany) and also in Utah (USA). Zippeite is no longer used for the manufacture of paints. It is used to obtain radioactive elements, as is pitchblende.

240 Zippeite — Jáchymov (Czechoslovakia); 10 × 6 cm.

**Scheelite** (241) was often found in cassiterite mines in the Erzgebirge and was called 'the white bronze hailstone' (in contrast to the so-called 'black hailstone' of cassiterite). Axel Cronstedt, a Swedish metallurgist and chemist gave a more accurate description of this mineral by the middle of the 18th century. He considered the mineral to be an iron ore and named it 'tungsten' (heavy stone). A detailed description and a chemical analysis of the mineral came from his countryman K. W. Scheele in 1781, who discovered the element tungsten in it. The mineral was later named scheelite in his honour by the French mineralogist F. Beudant (1781—1850).

Calcium tungstate, $CaWO_4$, tetragonal. H. 4.5–5; Sp. gr. 5.9–6.1; commonly greyish-white, yellowish, brownish, brown to reddish, rarely yellow to greenish, extremely rarely colourless; greasy, often even adamantine lustre; S. white.

Scheelite crystallizes in bipyramids, often reniform. It is not frequently massive. It occurs fairly abundantly in cassiterite and wolfram deposits. It crystallizes from hot solutions, frequently under the influence of gases (pneumatolytically) and occurs in quartz veins in granites and gneisses. Crystallized specimens of scheelite are greatly valued and are a gem of any collection. It is found in Zinnwald (East Germany) and in the Bohemian Cínovec, where it forms tiny, pale brown crystals and druses in quartz together with zinnwaldite. Other occurrences are at Schwarzenberg (Karl-Marx-Stadt district, East Germany), Cornwall (England) and Connecticut (USA). It is recovered as an important ore of tungsten.

The mineral **stolzite** is also included in this group; it was named after J. A. Stolz, a doctor from Teplice, Bohemia, who was the first to describe it as an independent mineral. It is lead wolframate ($PbWO_4$), which forms tiny tetragonal rounded crystals in shape similar to scheelite. They are often grouped in sheaf-shaped nodular clusters. Small quantities of stolzite occur in quartz at Cínovec (northern Bohemia), from where it was first described.

241 Scheelite — Cínovec (Czechoslovakia); 11 × 4 cm.

242 Crocoite — Dundas (Tasmania); 7.5 × 7.5 cm.

**Crocoite** (242) is a comparatively rare mineral, but the first one in which the element chromium was discovered. This success was earned by the French chemist Louis Vauquelin in 1797 when experimenting on crocoite from the Urals. Crocoite forms beautiful, fiery-coloured crystals, columnar to needle-shaped. They occur in chromite deposits, quartz veins and in the oxidation zone of lead lodes.

Lead chromate, PbCrO$_4$, monoclinic. H. 2.5–3; Sp. gr. 5.9–6; orange-yellow, with greasy to adamantine lustre; S. orange.

The most important deposits are at Berezovskiy, near Sverdlovsk in the Urals and Dundas in Tasmania. It occurs less abundantly at Baita (Romania), at Labo (Luzon Island, Philippines), near Congonhas (Brazil), Vulture (Arizona, USA), and in the Penchalong mine of Rhodesia in Africa. As it is so uncommon, crocoite it not important as an ore of chromium.

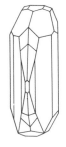

214

243 Wulfenite — Los Lamentos (Mexico); 6.5 × 6 cm.

**Wulfenite** (243), named after the Austrian mineralogist, A. Wulfen, who was the first to describe it in 1785, is a mineral of exceptional beauty, and also rather rare. It crystallizes in bipyramids, and is often tabular, and occasionally has short-columnar form. It occurs through oxidation of galena and therefore is mainly found in the oxidation zones of lead ores.

Lead molybdate, $PbMoO_4$, tetragonal. H. 3; Sp. gr. 6.7 to 6.9; commonly orange-yellow, orange, yellowish-grey to brownish, with adamantine to greasy lustre; S. white or light grey.

The main deposits: Bleiberg (Carinthia, Austria), Höllenthal (near Garmisch, West Germany), Mezica (Yugoslavia), Băita (Romania), Phoenixville (Pennsylvania, USA), Broken Hill (New South Wales, Australia). At Villa Ahumada in the Sierra de los Lamentos range in the Chihuahua province of Mexico there are occurrences of wulfenite of exceptional beauty. It develops in the form of rich clusters, made up of orange to brown, thick-tabular crystals with irregular edges. Crystals embedded in parent rock are very poor in faces and their average size is 2 to 4 cm. When it occurs in large enough quantities, wulfenite can be recovered as lead ore. Pure wulfenite contains up to 55 per cent lead.

## Chapter 8

# PHOSPHATES AND SIMILAR COMPOUNDS

This group, which contains not only phosphates and salts of phosphoric acid, but also arsenates and the rare vanadates, is second only to silicates, as far as the numbers of varieties are concerned. Here, too, we distinguish between the anhydrous salts (e.g. apatite, pyromorphite), the hydrous varieties (e.g. vivianite, variscite), and the alkaline salts (e.g. amblygonite, lazulite and others). The minerals in the phosphate group usually display isomorphism, where not only metals, but also phosphorus, arsenic and vanadium may substitute for each other. In many alkaline salts hydroxyl is substituted by halogen elements (as for instance in amblygonite). An interesting and distinct isomorphous group is represented by uranium mica, thus called because of its crystal shape and perfect cleavage. A large isomorphous group is composed of minerals — apatite, pyromorphite, mimetesite and others, which mix together. Only rarely does one come across a mineral which is a phosphate or a similar compound and which does not mix with other minerals (such as amblygonite and lazulite).

Phosphates usually occur in pegmatites and as a product of the alteration of other minerals. Most of the arsenates and vanadates develop secondarily on ore deposits.

**Amblygonite** (244) received its name from the Greek 'amblygonios' — slant-angled, because of its excellent cleavage and the angle of its cleavable planes. Most commonly it occurs in the form of irregular grains in pegmatites, often in association with lepidolite and tourmaline (rubellite). Its occurrence is comparatively rare. The main deposits are: Penig (Karl-Marx-Stadt district, East Germany); Vernéřov and Aš (Bohemia); Kashmir; Pala (California, USA); East Hebron (Maine, USA); São Paolo and Minas Gerais (Brazil). If fairly abundant, amblygonite is recovered as a lithium mineral (it contains up to 8.8 per cent lithium). When pure it is used as a semiprecious stone. The most suitable for this purpose are the unusual bladed amblygonite crystals recently discovered in the Minas Gerais state of Brazil. They are yellow and commonly up to 15 cm long. Practically all the yellow faceted cuts of this rare precious stone today originate from this particular deposit, which produces ample quantities of the beautiful, pale-coloured mineral.

Lithium-aluminium phosphate with fluorine and hydroxyl, $LiAl(F,OH)PO_4$, triclinic. H. 6; Sp. gr. 3–3.1; white, light green, yellowish, pink; vitreous lustre, pearly on some planes; S. white.

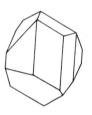

244 Amblygonite — Penig (German Democratic Republic); 7 × 4 cm.

**Lazulite** (245) was a familiar decorative stone in the past, but it was never highly valued. It was mainly used as an imitation of more precious stones. The Arabs, who excelled in the art of imitation of precious stones, passed it off as turquoise or the more valuable lazurite (lapis lazuli). This is why it is sometimes called 'imitation lapis'. Its modern name comes from the medieval Latin 'lazulum' — lapis lazuli.

Basic aluminium-magnesium phosphate with iron, $(Mg,Fe)Al_2(OHPO)_2$, monoclinic – pseudorhombic. H. 5 to 6; Sp. gr. 3.0; blue, with vitreous lustre; S. colourless.

Lazulite occurs in pegmatites, in quartz veins and in metamorphic rocks rich in aluminium. It is usually granular to compact, and only rarely forms sharp needle-shaped or tabular crystals. They seemingly have orthorhombic symmetry and commonly are twinned according to two different laws. Main deposits are in Salzburg (Austria), the Crowders Mountains of North Carolina and at Graves Mountains, Lincoln County, Georgia (USA), and at Tijoca in Minas Gerais (Brazil). In Crowders Mountains it is found fairly abundantly in flexible micaceous quartzites. Graves Mountains are noted for the nicest crystal specimens. The dark blue crystals from Tijoca are the ones most suitable as gemstones. There are interesting occurrences of lazulite also in the smaller deposits of Europe, particularly at Krieglach (Styria, Austria), where it is found massive, Zermatt (Switzerland) and in Värmland (Sweden). Today it is ground to lenticular shape or used for the manufacture of smaller ornamental objects.

245 Lazulite — Stickelberg (Austria); 12 × 9 cm.

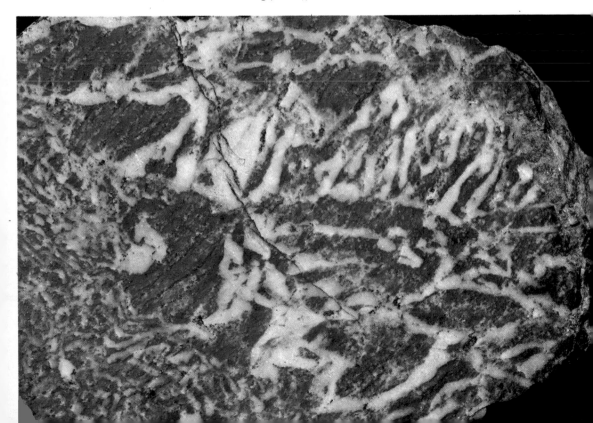

**Libethenite** (246) was found at the beginning of the 19th century when mining copper ores in the area of Lubietová near Banská Bystrica in Slovakia. Friedrich August Breithaupt, a German mineralogist, identified this mineral in 1823, described it and named it after the German name (Libethen) of the place where it was discovered.

Basic phosphate of copper, $Cu_2OHPO_4$, orthorhombic, H. 4; Sp. gr. approximately 3.8; dark olive-green, with greasy lustre. S. olive-green.

Libethenite commonly crystallizes in small crystals, nearly octahedral in form. It is dark olive, more rarely dark emerald in colour while larger crystals are nearly black in colour. Most frequently it occurs in cavities of quartz in association with other secondary copper minerals.

Another arsenate similar to libethenite is **olivenite** $Cu_2OH\ AsO_4$, with which it also often occurs. It can be distinguished from other similar secondary minerals chiefly by the crooked faces of the short and elongated columnar crystals. Reniform to globular aggregates are more common. It can be distinguished from the similar mineral malachite by its green to olive-green colour. Apart from Lubietová, libethenite has been found in Cornwall, England, where there are very beautiful crystals. There are deposits at Vysokaya Gora (Urals, USSR), in Yerington (Nevada, USA) and in Japan. It occurs less abundantly in a few other localities — France, Romania, Shaba (Zaïre), Arizona, Pennsylvania. Libethenite has only mineralogical significance but is much sought after by collectors. Naturally enough, libethenite from the original deposits of Lubietová is valued more highly.

246 Libethenite — Lubietová (Czechoslovakia); actual size of detail 5 × 2.5 cm.

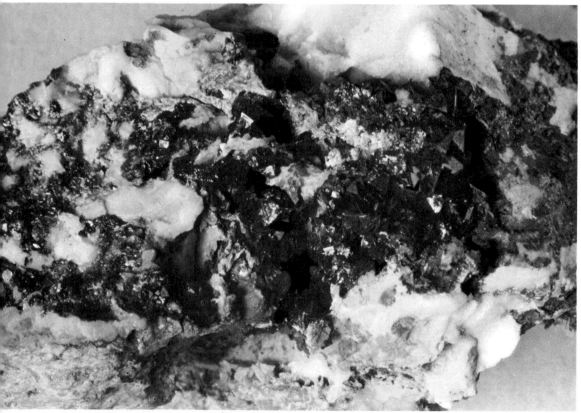

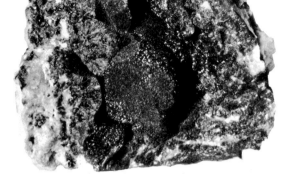

247 Pseudomalachite — Siberia (USSR); 12 × 9 cm.

248 Pseudomalachite — Rheinbreitenbach
(Federal Republic of Germany); 6.5 × 5.5 cm.

**Pseudomalachite** (247—248) was given its name because of its close resemblance to malachite.
It is also known under other names, for instance, **lunnite** in honour of
Lunn the chemist, **ehlite** because of the discovery of its deposits at
Ehle near Linz in the Rhineland (West Germany). It is also known as
**phosphorochalcite**. Not until fairly recently was it discovered that all
these minerals are identical.

Basic phosphate of copper,
$Cu_5(OHPO_4)_2$, monoclinic. H.
4.5; Sp. gr. 4.34; emerald to
blackish green, with dull vitreous lustre; S. green.

Pseudomalachite forms rich emerald-green reniform aggregates with
radiate-fibrous structure, or thin shapeless layers. It occurs as a secondary
mineral in copper ores, especially of chalcopyrite. Usually it is found in
association with malachite. Apart from Rhineland, the main deposits are
in Siberia, the surroundings of Chotěboř in Bohemia and Lubietová near
Banská Bystrica in Slovakia. Pseudomalachite so far has no practical
significance, for it never occurs in substantial quantities. It is only
extracted as a component which accompanies all the others in copper
ores.

A number of other minerals are included in this group. Let us at least
mention **tirolite**, named after the Tirol where it is found. Tirolite is
a hydrous basic arsenate of copper, and forms pearly orthorhombic
pseudohexagonal plates or leaves of blue-green colour. This mineral
originates from the decomposition of primary copper ores, particularly
near Falkenstein and Schwaz (Tirol, Austria), in Reichelsdorf in Thuringia and near Schneeberg (Karl-Marx-Stadt district, East Germany),
also in the Mammoth mine near Eureka, Utah (USA).

The rare **vesignieite**, a monoclinic basic vanadate of copper and
barium $BaCu_3(OH VO_4)_2$, forms tabular and powdery aggregates of
green to yellow-green colour. It occurs in the oxidation zone of hydrothermal ore-veins. It is found, for instance, in association with chalcedony
in the quartz veins in Vrančice (Bohemia).

219

249 Brazilianite — Minas Gerais (Brazil); crystals 5.5 × 4.5 cm and 6 × 3.5 cm.

**Brazilianite** (249) is one of the newly discovered minerals. It was identified by the American mineralogists Pough and Henderson in 1945 and named after the country where it was discovered. It occurs in the form of perfect crystals grouped in druses, in pegmatites, and is often of precious-stone quality. The only noted deposit of brazilianite is in the surroundings of Conselheiro, Pena, in the state of Minas Gerais in Brazil. During the past few years this deposit has yielded a great quantity of beautiful raw material, which has included crystals of surprisingly large dimensions and perfectly bounded crystal faces. Some of these are found on leaves of muscovite with their strong silvery glitter, ingrown in the parent rock. Such specimens of course are not ground, but find their way into museums and private collections. The most exquisite crystals, dark greenish-yellow to olive-green, sometimes measure up to 12 cm in length and 8 cm in width. While this book was being compiled, crystals of similar shape and dimensions were discovered in another deposit in Minas Gerais, near Mantena, but they lack the perfection of the crystal bounding. During the same period, brazilianites from yet another fresh locality appeared in many large collections; they were from the Palermo mine and the Charles Davis mine near North Grafton (Grafton County), New Hampshire, USA. It is a much sought after precious stone, usually ground into facet cuts, and it is a very popular item with collectors.

Basic aluminium-sodium phosphate, $NaAl_3[(OH)_2(PO_4)]_2$, monoclinic. H. 5.5; Sp.gr. 2.98; yellow, with vitreous lustre; S. white.

**Vanadinite** (250) forms columnar, rarely pyramidal crystals; fibrous aggregates with reniform surface; or it occurs massive. Vanadinite develops in the oxidation zone of lead ores. The main deposits are: Obir Mountain in Carinthia (Austria), Berezovskiy near Sverdlovsk in the Urals, Sierra de Cordoba in Argentina, Arizona, Zimapan in Mexico, Algeria, Morocco, Dumfries in Scotland and Australia. Vanadinite used to be an important ore of vanadium (it contains 19.4 per cent vanadium pentoxide). The alloy of vanadium and iron (ferro-vanadium) is used for the manufacture of special steels. Vanadium pentoxide is an important catalyst in the manufacture of sulphuric acid. Today, however, vanadium is gained mainly as a secondary product during the working of some iron ores (such as those from the Taberg Mountain, near Jönköping in Sweden and the Urals) and from bauxite.

Vanadate of lead with chlorine, $Pb_5Cl(VO_4)_3$, hexagonal. H. 3; Sp. gr. 6.8–7.1; commonly brownish, yellow to ruby red, with almost adamantine lustre, or strong vitreous lustre; S. white.

**Campylite** (251) received the name from the Greek 'kampylos' — bent, on account of the barrel-shaped bend of its crystals. It is composed of pyromorphite and mimetite (chloro-arsenate of lead), and usually occurs in association with them. It occurs in the upper lead deposits through the oxidization of galena or cerussite. The main deposits are Příbram in Bohemia and Dry Gill, Caldbeck, near Wigton, Cumbria, England. Campylite is a comparatively rare mineral.

Arsenate and phosphate of lead with chlorine, $Pb_5Cl[(As,P)O_4]_3$, hexagonal. H. 3.5–4; Sp. gr. 6.7–7.1; yellow to orange, adamantine to greasy lustre; S. whitish.

Another comparatively rare mineral is **mimetesite,** or **mimetite,** whose name is derived from the Greek 'mimetes' — the imitator, for it resembles in shape the crystals of pyromorphite, for which it used to be mistaken. Most commonly it forms hexagonal yellowish columns, usually in rounded barrel-shaped forms, which are typical of campylite. In this characteristic it differs from pyromorphite. The chief sources are Johanngeorgenstadt (East Germany), Nerchinsk (Siberia) and Tsumeb (Namibia). In America it is found in Mexico and at Phoenixville (Pennsylvania, USA).

250 Vanadinite — Djebel Mahser (Morocco); crystals 2—2.5 mm.

251 Campylite — Příbram (Czechoslovakia); aggregates 4—9 mm.

252 Pyromorphite — Příbram (Czechoslovakia); 9 × 8 cm.

**Pyromorphite** (252) has been known for a long time, though it has been called several names, according to its colour. The present name originated from the Greek 'pyr' — fire, and 'morph' — form, for it used to be wrongly thought that the crystals were the result of volcanic magma. Today it is known that, on the contrary, it is formed in the upper zone of lead veins in association with cerussite and a number of other secondary minerals, which originated through the decomposition of galena. It is by far the most conspicuous of all these minerals by its glaring colour and by the shape of its crystals. They are usually columnar, often perfectly bounded by smooth faces and grouped into druses. On rare occasions it forms reniform or botryoidal aggregates. The conspicuously green colouring influenced its earlier German name 'Grünbleierz', whereas when pyromorphite was found in the less common brown variety, it was classified as 'Braunbleierz'. Pyromorphite with an admixture of calcium is called **miesite**. This name was given from the German name of the locality in Bohemia where it occurred — Stříbro (Mies). Here it develops whitish radial aggregates in association with brown pyromorphite.

Phosphate of lead with chlorine, $Pb_5Cl(PO_4)_3$, hexagonal. H. 3.5 to 4; Sp. gr. 6.7–7.0; commonly green or brown, rarely yellow, orange, white to colourless; adamantine to greasy lustre; S. whitish.

 The main deposits are: Clausthal-Zellerfeld (Harz Mountains, West Germany), Johanngeorgenstadt (Karl-Marx-Stadt district, East Germany), Cornwall (England), Leadhills (Scotland), Příbram (Bohemia), Banská Štiavnica (Slovakia), Berezovskiy in the Urals and Nerchinsk (USSR), Phoenixville in Pennsylvania and Mullan in Idaho (USA), Brazzaville (Congo) and Broken Hill (New South Wales, Australia). It occurs in association with other lead ores, and if found in large enough quantities, serves as a lead ore with a content of 75 per cent lead.

**Apatite** (253—256) received the name from the Greek 'apate' — deceit. Ancient mineralogists mistakenly thought that every mineral had its own characteristic colour. They took apatite for many varied minerals, such as aquamarine, amethyst or tourmaline. When, at the end of the 18th century, its chemical composition was determined and it was established as a distinct mineral, it was aptly named apatite, for its misleading appearance, by Abraham Gottlob Werner, professor at the Mining Academy in Freiberg.

Phosphate of calcium with fluorine and chlorine, $Ca_5(F,Cl)(PO_4)_3$, hexagonal. H. 5; Sp. gr. 3.16–3.22; commonly blue-green, green, grey, white, brown, with greasy lustre; S. white.

Apatite occurs in a variety of colours, and also in a variety of forms. Its crystals can be thickly tabular, columnar, elongated to needle-shaped. It also occurs in massive, granular, radial and compact forms. Apatite is present in most rocks, especially in pegmatites and cassiterite deposits, but not in large quantities. There are, however, sedimentary rocks which often form whole layers and whose basic component happens to be apatite. Such rocks are called phosphate rocks (phosphorite) (253). Such apatites are not completely chemically pure, for they contain a number

253 Apatite — phosphorite — Podoli (Ukraine, USSR); 7 × 6 cm.

223

254 Apatite — Greifenstein (German Democratic Republic); crystals up to 15 mm.

255 Apatite — Durango (Mexico); crystals approx. 1.5 cm.

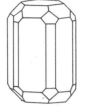

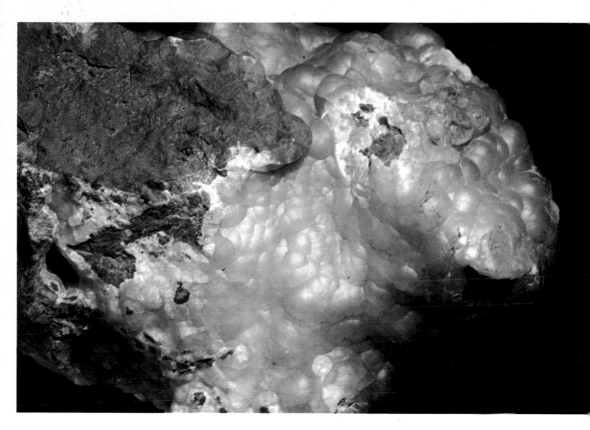

256 Apatite — staffelite — Staffel (Federal Republic of Germany); actual size of detail 11 × 8 cm.

of various impurities and organic remains. Commonly they are compact, fibrous or reniform, with a wide range of different shades of grey. Their origin is explained by several theories, which often contradict each other. In general, one can say that organisms take compounds of phosphorus from seawater and accumulate them in the hard parts of their bodies. After their death, the gradually released phosphate is deposited often in colloidal form, particularly in poorly aerated waters. **Staffelite** (256) is a stalactitic apatite and occurs in the Staffel deposits near Limburg, West Germany.

The largest deposits of phosphates (i.e. apatites used for commercial purposes) are in the USA (Florida and Tennessee), yielding almost one-half of the world's production of phosphates. Further occurrences are in North Africa (Tunisia, Morocco, Algeria and Egypt), in the USSR (Kazakhstan, Ukraine and the Kola Peninsula). The best known apatite crystals come from St Gotthard (Switzerland), Saxony, Quebec (Canada) and Durango (Mexico).

Phosphates are an excellent source for the manufacture of fertilizers. The attractively coloured apatite varieties are used as ornamental stones. The brilliant cuts of the pink and yellow Swiss apatites are well known, and quite beautiful. Understandably, they are also rather expensive. The yellow-green apatites are also very popular as semiprecious stones; they are called 'asparagus stones' because their colour resembles that of asparagus.

225

257 Vivianite — Košťálov (Czechoslovakia); 4 × 3 cm.

**Vivianite** (257) was discovered at the beginning of the 18th century by the British mineralogist J. G. Vivian in pyrrhotite veins near St Agnes, Cornwall. In 1817 Abraham Gottlob Werner named the mineral in his honour. Later vivianite was discovered in substantial quantities in other deposits, first in massive form, later also in crystal form.

Hydrated iron phosphate, $Fe_3(PO_4)_2 . 8H_2O$, monoclinic. H. 2; Sp. gr. 2.6–2.7; light to dark blue, to greenish, with vitreous lustre, pearly on some planes, with metallic shades; S. colourless to blue-white, which quickly turns to indigo-blue.

Vivianite crystals are commonly columnar or tabular, often in star-shaped or round-shaped groups. It is also earthy or powdery (in peat bogs). When fresh it is clear, but with oxidation turns immediately blue or green. This occurs because the bivalent iron in the vivianite partially changes into trivalent iron. The crystal varieties are noted for their dark blue colour, but when they are in powdery form, it is a pale blue. The fine vivianite leaves are flexible.

Vivianite develops in pyrite deposits, mostly through their weathering, and in clay, mud and peat-bog iron ore, also in coprolites, fossil shells and bones. Here it usually forms star-shaped aggregates together with gypsum. The familiar 'blue clay' in peat is powdery vivianite which developed under normal temperature through the action of loose phosphoric acid of organic origin upon limonite. Vivianite is frequently associated with limonite, and vivianite crystals are also commonly found in limonite cavities.

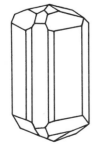

The main deposits apart from Truro and St Agnes, Cornwall, are: Bodenmais in Bavaria (West Germany), Bohemia, Kerch (USSR), Cameroon, Mullica Hills in New Jersey and Leadville in Colorado (USA).

Vivianite is of no commercial value so far, but the well developed crystals are popular with collectors. This applies especially to the enormous crystals of vivianite found in Cameroon.

226

**Bukovskyite** (258) is today one of the world's acknowledged mineral rarities, though it used to be collected a long time ago from overgrown pit heaps by the inhabitants of Kutná Hora in Bohemia. It was used for poisoning fieldmice and other field vermin. This poisonous clay, known also by the place name of its discovery, 'clay of Kutná Hora', was widely known and it was considered to be 'arsenic' (arsenic trioxide). In 1901 Antonín Bukovský, a Bohemian chemist, who studied minerals of old pit heaps, proved it was an arsenate.

Hydrated basic arsenate and iron sulphate, $Fe_2SO_4AsO_4OH \cdot 7H_2O$, monoclinic. H. 3; Sp. gr. 2.33; pale yellow-green to grey-green, earthy; S. yellow-white.

Bukovskyite forms nodules with a reniform surface, which under a microscope appear as a collection of minute needles similar to gypsum. Some can be seen with the naked eye and occur inside the nodules. Bukovskyite has so far only been found in one locality of the world, on pit heaps from the Middle Ages, where sulphate ores had been mined at Kaňk, north of Kutná Hora and other old deposits in the vicinity. As it has only recently been defined and acknowledged, it has attracted great attention from mineralogists and collectors. This is because it is one of the few newly discovered minerals which occur also in fairly large, and fairly conspicuous pieces. One must handle it with care, however, for it is a highly poisonous mineral.

258 Bukovskyite — Kaňk (Czechoslovakia); 13 × 8 cm.

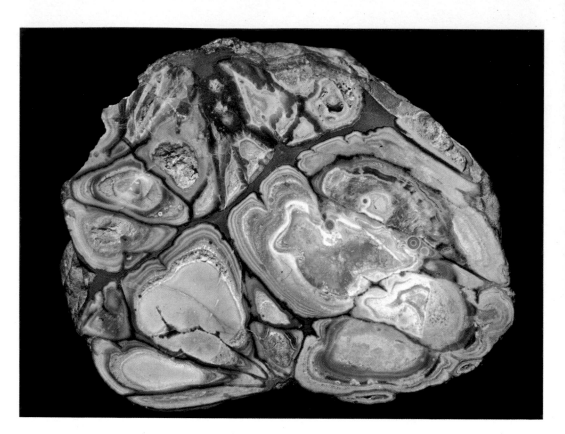

259 Variscite — Lewiston (Utah, USA); 10 cm.

**Variscite** (259) is a most attractive, but little known precious stone, which is often mistaken for other minerals. Variscite was described in 1837 by Friedrich Adolf Breithaupt, a German mineralogist from Vogtland, a region centred on Plauen in East Germany. Variscite was named after the old name of this region, Variscia. Crystals of the same mineral were found in 1894 on the western shore of Lake Utah in the USA, and were named **utahlite.** Later it was found they were the same mineral. Now the name utahlite is applied mainly to the semiprecious varieties of variscite. A variety rich in iron is called **redondite.**

Hydrated aluminium phosphate, $AlPO_4 . 2H_2O$, orthorhombic – pseudohexagonal. H. 4–5; Sp. gr. 2.52; bluish-green, white to colourless, with waxy lustre; S. colourless.

Variscite occurs in the form of radially fibrous aggregates and crusts with reniform surfaces, and occasionally in indistinct columnar crystals. It fills cracks and cavities of rocks rich in aluminium, mainly in silicate schists and greywackes, often in association with amorphous aluminium phosphates. The main deposits: Utah, Arkansas (USA), Plauen (East Germany), Thuringia (East Germany) and the Slovakian Ore Mountains. The loveliest specimens of variscite are egg-shaped to spherical tubers up to 30 cm long, which are found in breccia cavities of sedimentary rocks near Fairfield, Utah. Most of these concretions do not, however, exceed the average size of 7 to 10 cm, and many are suitable only to be polished into a plate, but not be used as a gemstone. Variscite, whether on its own, or in association with wardite, is an extremely pretty, but little known precious stone, which is often mistaken for other minerals. It is commonly ground to lenticular shape.

228

**Turquoise** (260—261) was a most popular precious stone, especially during the time of the Otto-man Empire. The Turks bought the stone and transported it to Persia. Apart from the Ottoman Empire, there were also turquoise mines on the Sinai Peninsula, which were worked in the days of ancient Egypt, but were in the hands of wild tribes, which were hostile towards the Turks. The Turks, though aggressive, were forced to pay dearly to the local tribes for the turquoise, and for some time turquoise was an extremely expensive precious stone. There were also smaller deposits at that time in Anatolia.

Hydrated basic phosphate of aluminium and copper, $CuAl_6[(OH)_2PO_4)]_4 . 4H_2O$, tri-clinic. H. 5–6; Sp. gr. 2.6–2.8; commonly green-blue, blue to grey-green, with a faint waxy lustre; S. white.

Turquoise is cryptocrystalline. It is commonly reniform, stalactitic, or forms a coating. It occurs as a filling in cracks in siliceous schists and sandstones, occasionally in pegmatites. The main deposits are: the vicinity of Nishapur in Iran; Karatobe in Kazakhstan (USSR); Los Cerillos Mountains in New Mexico and in Utah and California (USA). In Europe turquoise occurs in Jordanów in Silesia (Poland). The attrac-tively coloured varieties are used as precious stones. The beautifully crystallized variety is found only in a small mine for copper ores near Lynch Station, Virginia. The occurrences in the form of thin crusts and veins which permeate rocks are in practice the most important. The pseudomorphs after felspar, which come from Nishapur in Iran, are famous from the mineralogical and collector's point of view; the pseudo-morphs after apatite found in California are also well known. The richly blue turquoise is the most valued variety; it is fairly rare and gener-ally permeated with brown stains from iron compounds. The colour of some turquoise varieties is not constant when exposed to light and shows little resistance to chemical influences, particularly to the effects of acids. Such disturbed turquoises temporarily regain their original colour if immersed in ammonia.

Turquoise is also produced artificially, but more often it is imitated by the so-called 'bone turquoise', which is in fact fossil bones tainted blue, or by cheaper stones such as lazulite.

260 Turquoise — Anatolia (Turkey); 4 × 3 cm.

261 Turquoise, cut — Iran; 3 × 2.5 cm.

262 Wavellite — Třenice (Czechoslovakia); 6 × 5 cm.

**Wavellite** (262) was discovered in 1800 near Barnstaple in Devon (England) and identified by the English physicist W. Wavell. It was found in larger quantities at the beginning of the 19th century near Hořovice, Bohemia, when mining iron ores and quarrying for building stone.

Hydrous basic phosphate of aluminium, $Al_3(OH)_3(PO_4)_2$. $5H_2O$, orthorhombic. H. 3.5 to 4; Sp. gr. 2.3–2.4; colourless, whitish to yellow-green, sometimes green, blue to brown, with vitreous lustre; S. white.

Wavellite forms radiate bunches, round or more rarely botryoidal or reniform aggregates, consisting of finely fibrous to needle-shaped small crystals. It occurs in sedimentary rocks, such as greywacke and sandstone. There it was formed by alteration and crystallization from dissolved shells of various animals. The best known deposits apart from Bohemia and Cornwall are: Langenstriegis near Freiberg, Karl-Marx-Stadt district, East Germany; the world-famous magnetite mines near Kiruna in north Sweden; Chester in Pennsylvania, USA; and Ouro Prêto in Brazil. Wavellite has hardly any practical significance. In some places in the USA, however, it occurs in substantial quantities and is recovered as a raw material for obtaining phosphorus, which was previously mainly used for the manufacture of matches. Today paint is still made from it in Harrisburg (Pennsylvania). Though wavellite is a fairly common mineral, it is much sought after by collectors on account of its conspicuously radiating bunches. Lumps with wavellite in globular form and with the relatively abundant **zepharowichite** (microscopically compact wavellite) are particularly popular.

**Torbernite** (263), **autunite** and **carnotite** belong to the uranium mica group. Miners found torbernite and autunite long ago in the ore-veins of the Erzgebirge (Ore Mountains) in Saxony and in neighbouring Bohemia. They named them 'the green mica'. The first scientific reference to the minerals was given by Ignác Born (1742—1791), a professor at Prague University, and by Abraham Gottlob Werner (1749—1817), a mineralogist and a geologist from Freiberg, at the end of the 18th century. But it was not until 1797 that a famous German chemist, M. H. Klaproth discovered that these two minerals, in contrast to all the other known normal mica varieties, contained a new element, uranium. It was also discovered that apart from being green, torbernite and autunite could also be yellow-green or yellow in colour.

Torbernite and autunite most commonly crystallize in thin tabular plates, often well bounded, and are noted for their excellent cleavage. They occur more abundantly as flaky aggregates or bunches on the walls of cracks.

Torbernite (named after the Swedish chemist Torbern Bergman) or **chalcolite** (from the Greek 'chalkos' — copper, and 'lithos' — stone) and autunite (after the uranium ore deposit at Autun in France), are much alike in shape, but diverse in colour. These minerals originate from the weathering of uranium. The main deposits are: Cornwall in Britain, Tannenbaum gallery in Johanngeorgenstadt (Karl-Marx-Stadt district, East Germany), Jáchymov in Bohemia, Brancheville in Connecticut (USA), Shinkolobwe in Zaïre, Watsonville in Australia. The main deposits of carnotite are in Utah, USA. Both these minerals are important for the production of uranium and its compounds. Autunite contains 60 per cent $P_2O_8$, torbernite 55 to 60 per cent.

Hydrated phosphate of uranium and copper, $Cu(UO_2PO_4)_2 . 8-12H_2O$, tetragonal. H. 2–2.5; Sp. gr. 3.3; emerald green, with pearly lustre; S. apple-green.

Autunite:
Hydrated phosphate of uranium and calcium, $Ca(UO_4 . PO_4)_2 . 8-12H_2O$, tetragonal. H. 2–2.5; Sp. gr. 3.2; yellow-green, sulphuric yellow to golden-yellow, with intensively vitreous lustre, often pearly on some faces; S. yellow.

Carnotite:
Hydrous vanadate of uranium and potassium, orthorhombic. H. 4; Sp. gr. 4.46; yellow to yellow-green, pearly lustre; S. yellow-grey.

263 Torbernite — Horní Slavkov (Czechoslovakia); 7 × 5 cm.

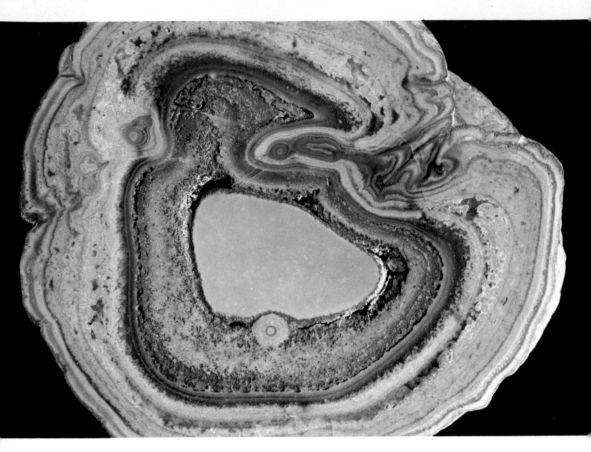

264 Wardite — Livingstone (Utah, USA); 6 cm.

**Wardite** (264) is one of the least known precious stones. It was first defined by the American mineralogist, Davis, after it was found in Cedar Valley near Lake Utah, USA, so far the only place it has been found. It appears in small quantities in association with variscite utahlite, on whose nodules it forms oolite in compact layers. As a precious stone it comes into consideration only when together with variscite. It developed in this deposit in underground layers from the effects of the gradually infiltrating ground waters containing dissolved phosphates, and from their influence on rocks rich in aluminium. A number of various phosphates and other minerals develop with it. Apart from variscite, these are chiefly wavellite, millisite, gordonite, crandallite, the internally crystalline apatite, chalcedony, goethite and clinovariscite. The above mentioned **millisite** is a hydrous basic phosphate of aluminium, calcium and sodium, which forms grey fibrous concretions. **Gordonite**, triclinic hydrous basic aluminium-magnesium phosphate, usually occurs in colourless layers, and only rarely in crystallized form. The earthy-yellow **crandallite** is a hydrous basic aluminium-calcium phosphate. **Clinovariscite** (or **metavariscite**) resembles variscite in appearance and composition, but is monoclinic. The 'eye-like' forms of these phosphates, formed by concentric layers, are particularly popular. Such specimens are polished as ornamental stones.

Complex hydrated basic phosphate of aluminium and sodium; Na Al$_{12}$(OH)$_{18}$(PO$_4$)$_8$ . 8H$_2$O, tetragonal. H. 5; Sp. gr. 2.8; pale green to blue-green, with waxy lustre; S. colourless.

# Chapter 9 SILICATES

This is the largest and commonest group of minerals, for it represents approximately 40 per cent of all minerals. In some silicates aluminium replaces silicon. Such minerals are then called aluminosilicates (felspar, kaolinite, for instance). Detailed systematization of silicates is rather complicated and is determined by the grouping of the internal units in the structure of the mineral. Some silicates, which resemble each other in their chemical structure, often intermingle and therefore form natural groups. The minerals of such groups share similar characteristics (garnets, pyroxenes, felspar, etc.). Zeolites form an unusual group of silicates — crystalline hydrated aluminosilicates of light metals. They differ from other crystalline minerals in retaining their internal structure and crystal form even with the evaporating of their water.

Silicates occur in nature in a variety of ways, as primary or secondary minerals in igneous or metamorphic rocks.

**Phenakite** (265) was found long ago in the well-known emerald deposits east of Sverdlovsk in the Urals, but it was mistaken for quartz. Count Petrovski, a Russian mineral collector, took away a few specimens while inspecting the mines, and upon closer examination it became evident that this was an entirely different mineral.

Beryllium silicate, $Be_2SiO_4$, rhombohedral. H. 8; Sp. gr. 3.0; commonly colourless, rarely wine-yellow to rosy, with vitreous lustre; S. white.

Phenakite occurs relatively rarely in the form of columnar crystals, chiefly in pegmatites. The crystals are sometimes twinned and appear to be hexagonally symmetrical, much like the crystals of quartz. This is why it took so long for this minerals to be accurately described. Apart from the Urals, the best known deposits are Kragerö in Norway, Pike's Peak (Colorado) in the USA and San Miguel in Brazil. It is used as a precious stone.

265 Phenakite — Kragerö (Norway); large crystal 20 cm.

**Garnets** (266—277) were named 'garnata' in the 13th century by a famous German theologist and philosopher, Albertus Magnus (1193—1280), perhaps from the Latin 'malum granatum' garnet apple, in accordance with their most frequent colour, or perhaps from 'granum' — grain, referring to their shape. The garnet family also contains many minerals which in ancient times used to be called 'carbuncles', from the Latin 'carbunculus' — small red-hot coal. It is quite certain that Pliny the Elder used this term for the garnet almandine for instance.

Binary silicates of bivalent and trivalent elements, cubic. Vitreous to greasy lustre; S. white.

The garnet group comprises minerals of similar composition and structure, which mix with one another. They come in a greater variety of colours than any other minerals. This is because the group consists of a large number of binary silicates of bivalent and trivalent elements, which share isometric symmetry. They crystallize most frequently in the cubic system as rhombododecahedra or trapezohedra. Very often they form granular to massive clusters in the rocks. They have no cleavage, but often distinct partings. As garnets of different composition intermingle, a pure variety hardly ever exists. Garnets are among the most widespread minerals.

266 Garnet — pyrope — Central Bohemian Highlands (Czechoslovakia); diameter of the bored rock segment 11 cm.

267 Garnet — almandine — Starkoč (Czechoslovakia); 7 × 6.5 cm.

In the mineral kingdom garnets develop under the most varied conditions, but always under high temperatures. Perfectly bounded crystals form in gneisses and mica schists, but also in phyllite, marble, serpentines and granite pegmatites. As garnets are resistant to weathering, they often occur in alluvial deposits. Garnets often occur in contact rocks, which develop through contact of igneous rocks with other rocks. This is the origin of garnet rocks, contact rocks formed chiefly of garnets, pyroxenes and minerals with magnesium.

Pyrope:
Magnesium-aluminium silicate, $Mg_3Al_2(SiO_4)_3$. H. 7–7.5; Sp. gr. 3.7-3.8; commonly red.

**Pyrope** (266) is undoubtedly the most popular of the garnet family. The most important deposits occur on the southern slopes of the Bohemian Highlands, where they have been found since the end of the 16th century, and this was when the largest Bohemian garnet was discovered. It was the size of a pigeon's egg and thus exceptionally rare. Pyropes which are even the size of a pea are very rare.

All the occurrences of the garnet are described by Anselmus Boëtius de Boot in his book *Gemmarum et lapidum historia*. De Boot writes not only about the garnet the size of a pigeon's egg, but about other large

235

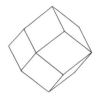

268 Garnet — almandine — Přibyslavice (Czechoslovakia); crystals 3 cm.

pyropes. The book informs us, for instance, that pyropes of the size of a hazelnut were valued as highly as rubies.

The popularity of the Bohemian garnet reached its height in the second half of the 19th century, when pyrope jewelry was exported chiefly to Poland and Russia; they were in great demand at the court of the Tsar.

The parent rocks of pyrope are chiefly serpentines and rocks rich in olivine, but the most common occurrence is in alluvials. Pyrope usually forms dark, fiery red grains of vitreous lustre. This is also the origin of its name, which is derived from the Greek 'pyropos' — fiery-eyed. The colouring is caused by the admixture of chromium and iron.

269 Garnet — almandine — Sederalpe (Tirol, Austria); crystal 4 cm.

270 Garnet — almandine — Salida (Colorado, USA); 6.5 cm.

271 Garnet — hessonite — Žulová (Czechoslovakia); actual size of detail 16 cm.

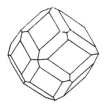

The most noted deposits are, apart from the Bohemian Highlands, the diamond mines near Kimberley, South Africa, in Arizona, Utah and New Mexico, USA and in kimberlites on the diamond deposits at Yakutsk (USSR). Garnet is often named after the locality of its occurrence, such as the Bohemian garnet, Arizonian ruby, etc. Lesser known deposits are also found in Australia, Tanzania, Rhodesia and Brazil. The light rose-red to faintly violet mixtures of pyrope and almandine are called rhodolites and they occur in California (USA).

Bohemian garnets (7) are today much sought after as gemstones. Their popularity is caused by their beautiful colouring and lustre, which are

272 Garnet — hessonite —
Ala (Piedmont, Italy);
crystals 5 mm.

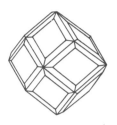

Almandine:
Iron-aluminium silicate,
$Fe_3Al_2(SiO_4)_2$. H. 7; Sp. gr. 4.1
to 4.3; commonly red.

further enhanced by cutting and polishing. However, Bohemian garnets are rarely very large. 'Cape rubies', named after the locality of their discovery, are much larger in size.

Garnets which have the almandine composition are the most abundant. They are also the chief component of the so-called common garnet. **Almandine** (267—270) received its name from the location of its first discovery — the surroundings of the ancient city of Alabanda in what is now southwest Turkey, where it was mined in the past. At that time almandines with a violet hue were particularly valued. Among all garnets, almandine is the one which has been used the longest as a precious stone (2).

Schists are the most common parent rocks of almandines. Almandine crystals are usually fairly large (in comparison to pyrope for example) and frequently perfectly bounded. They often retain their high gem

273 Garnet — hessonite — Žulová (Czechoslovakia); crystals up to 16 mm.

274 Garnet — spessartite — Tokovaya river (Urals, USSR); crystals 8—9 mm.

quality even when they are large. Because of this there are quite a few well known cut almandines, exceptional in size and perfectly coloured. The largest stone is in the collection of the Smithsonian Institution in Washington. This is a star (asteric) almandine from the state of Idaho, USA, and weighs 175 carats. Asterism is an optical effect which occurs through the reflection of light in the direction of the crystal axis. Rays of light transmitted in this direction fall on microscopic inclusions in the crystal which are arranged parallel to the crystal axis. When viewed from above, a distinct star can be seen on the mineral.

Almandine resembles pyrope in colour, but it is distinguishable by its violet or brownish hue. It is also usually paler than pyrope. The different hues are the result of different admixtures. Pure almandine is an exception. More frequently it contains some pyrope components, some spessartite, and often further admixtures. A violet hue, for instance, is caused by the presence of iron and chromium.

The most important almandine deposits of gemstone quality today occur in Sri Lanka, India, Australia, Madagascar, Brazil and the USA. The most noted European localities are in the Austrian Tirol, in Bohemia near Čáslav and in Romania. Almandine's value as a precious stone is

239

today related to the degree of resemblance to the ruby and the best come chiefly from the gemstone alluvials in Sri Lanka (the so-called 'Ceylon rubies'). In India, where there are numerous almandine deposits, large, rounded crystals are found in alluvials. The Tirolese almandine crystals, intergrown into the grey-green slate, are frequently beautiful, perfectly bounded crystals measuring as much as 5 cm or even more. Fairly recently, a new deposit of attractive almandine crystals has been discovered in Alaska, where they are intergrown into dark grey mica-schists.

**Andradite**, the next mineral in the garnet group, received its name from the Greek 'andras', used originally by a Greek philosopher, surgeon and natural scientist, Theophrastus (372—278 BC). At that time he applied the name to all the garnet family. Andradite is commonly massive to compact. It often occurs in schists and in seams, and is frequently found in association with pyroxene (hedenbergite) or amphibole; it also occurs in serpentines. The gemstone varieties, mainly demantoid and melanite (277) are the most important.

The green **demantoid** (277) is without doubt the most valuable garnet. It was first discovered near Nizhniy Tagil in the Urals in the 1860s in gold-bearing alluvials. But it was found only 20 years later in its parent rock by the banks of the Bobrovsk in the Sysertsk region (USSR). Demantoid outshines all the other garnets in lustre and dispersive power — which is the ability to break up white light into the colours of the spectrum — and in this it is similar to a diamond. It was named accordingly from the obsolete German 'Demant' — diamond. Its alternative name — **topazolite** — is applied because the mineral resembles a topaz in colour. The emerald, light green or yellow-green colouring of demantoid is caused by the presence of chromium. The most recent occurrences of demantoids of gemstone quality come from Val Malenca in northern Italy and from Tanzania.

**Melanite** (277) owes its name to the Greek 'melas' — black. It is darkly brown to black, which is caused by the presence of titanium and sodium. Melanite usually occurs in crystal form, in volcanic rocks, such as phonolites. The main deposits are at Frascati near Rome and Vesuvius, Italy and also in West Germany. It is generally used in the manufacture of jewelry for mourning purposes.

**Grossular** is a garnet which was named after the scientific name for a gooseberry — *Ribes grossularia* — because of its gooseberry colour. It occurs compact and crystallized, and forms under high temperatures on the points of contact of granite magma rich in silicon dioxide and carbonate rocks; it is also formed through the action of limestone on rocks rich in aluminium. The main deposits are; the basin of the river Vilyuy and the Urals in the USSR. In several localities in South Africa it is found in a variety of colours (in Wollrütterskop for example).

**Hessonite** (271—273) or **essonite**, is basically a variety of grossular rich in iron. The name comes from the Greek 'hesson' — inferior, for it is of less value than zircon, which it resembles. In the chief deposits of hessonite's gemstone varieties, the placers of Sri Lanka, it occurs in association with a zircon variety for which it used to be mistaken. Hessonite is indeed almost indistinguishable from zircon in its hyacinth-red to orange colour.

Like grossular, hessonite is a typical contact mineral, which occurs at the points of contact of igneous rocks and limestones. This is why it is accompanied in its primary deposits by other contact minerals, chiefly wollastonite (white fibrous calcium silicate — $CaSiO_3$), pyroxene (diopside) and vesuvianite. Large crystals of hessonite occur near Žulová in the Jeseníky Mountains, Silesia (Czechoslovakia), also in Piedmont in northern Italy, on the banks of the river Vilyuy in the Yakutsk region (USSR), in Xalostoc, Mexico and Mudgee in New South Wales (Australia).

Andradite:
Calcium-iron silicate, $Ca_3Fe_2(SiO_4)_3$. H. 6.6–7.5; Sp. gr. 3.7–4.1; commonly greenish.

Grossular:
Calcium-aluminium silicate $Ca_3Al_2(SiO_4)_3$. H. 6.5–7.5; Sp. gr. 3.7–4.3; commonly green or reddish-brown (hessonite).

Another garnet — **spessartite** (274) — is not as well known as the others already mentioned. It was named after the locality of its first discovery — the Spessart Mountains between Frankfurt and Würzburg, West Germany. It nearly always contains a small or large admixture of almandine, so in colour it is extremely changeable. Usually it is yellow, orange, reddish, or brown; effects of weathering turn it dark to black. Spessartite usually occurs in crystal form.

Found chiefly in granite pegmatites and granites, and in other igneous rocks, its main deposits, apart from Spessart, are in the Harz Mountains, West Germany, at Miass in the Urals (USSR), Haddam in Connecticut (USA), Ampandramaika in Madagascar, and Sri Lanka. The transparent spessartite variety is a very popular semiprecious stone.

**Uvarovite** is considered to be the most beautiful green precious stone. Found mainly in chromium deposits, it is rather a rare garnet. The most beautiful specimens of uvarovite come from serpentines near Sysert and Saranaya in the Sverdlovsk region of the Urals, where it occurs in association with demantoid. Other deposits are at Jordanów in Polish Silesia, the Texas mine in Pennsylvania (USA), and Makri (Turkey). Uvarovite crystals are so minute that they are hardly ever suitable for cutting and polishing.

The non-transparent combinations of garnet compounds are called the **common garnet.** The colour differs according to chemical composition. The main component is usually almandine or andradite, and the most common occurrence is in schists and gneisses.

275 Garnet — Oravita (Romania); crystals 1 cm.

276 Garnet — Baia Sprie (Romania); 6 × 4.5 cm.

Only the most common garnets are described here. There are many more varieties. Garnets even occur which contain yttrium, vanadium and zirconium.

As precious stones, the most important garnet stones are pyropes, almandines and demantoids. The remaining ones have not the same significance for the gem industry. But many of them are also collected and worked. For instance, the reddish-orange hessonites from Sri Lanka deposits and from other areas are cut and polished as precious stones, the black melanites are used in the manufacture of jewelry for mourning purposes, and there are many others. Apart from being used as precious stones, garnets can also be of industrial value. Garnets are one of the best abrasive and polishing materials and they are used for the manufacture of abrasive paper. As an abrasive, garnet is superior to the harder quartz, for during use its shape is always sharp, which prolongs its abrasive efficiency. This occurs because garnets have a perfect fracture (not to be confused with cleavage). Garnet's cutting ability is therefore twice to six times greater than that of quartz sand. Garnet cloths and papers are used for grinding the most varied materials.

Their hardness makes garnets, especially pyropes, highly suitable as bearings in fine instruments. The physical properties of pyrope, especially heat conductivity and elasticity, are an advantage. Pyropes are particularly suitable stones for very accurate clocks.

Synthetic garnets, which have only been made in recent years, deserve special attention. Attempts at their manufacture started in the 1960s, at first for industrial uses. These synthetic garnets, which differ from the natural ones in their chemistry, have the highest refractive index and dispersive power of all synthetic stones. They are therefore particularly suitable as imitation diamonds. During the last few years they have appeared increasingly on the market, mainly under the name of YAG (yttrium-aluminium-garnet), and are of a high value. At present they are made to be either transparent, pink or light green.

277 Garnets, cut — from top to bottom: almandine — India; 3 pyropes — Central Bohemian Highlands (Czechoslovakia); spessartite — Madagascar; almandine — India; demantoid — Urals (USSR); melanite — Vesuvius (Italy); largest almandine 32.4 × 18.6 mm.

**Olivine** (278—279) was sought after as a precious stone in ancient Rome, and it was originally found on a small island which the Romans called Topazos. They named the beautiful green stone they found appropriately topazus and it is frequently mentioned in Latin works. Some time later, the study of these sources gave the impression, quite wrongly, that the Roman topazus was the stone now known as topaz. This large deposit was eventually forgotten, and it was thought that the Roman topazus came from India or somewhere in the Orient.

Silicate of magnesium and iron, $(Mg,Fe)_2SiO_4$, orthorhombic. H. 6.5–7; Sp. gr. on average 3.5; commonly yellow-green to olive-green; vitreous lustre; when fractured somewhat greasy; S. white.

Later, deposits of this stone were found in Europe. Anselm Boëtius de Boot writes how it was recovered on the hill of Kozákov in Bohemia. The mineral was called chrysolite, which means 'gold stone' in Greek.

Until 1900 Kozákov remained the sole known deposit of precious stone but then it was discovered that the island of Zebirget (known in English as St John Island) on the Egyptian shore of the Red Sea was in fact the lost island of Topazos. This rich deposit soon regained fame and today once again represents the main occurrence in the world, for Egyptian olivines are extremely pure and, more important still, considerable in size.

Olivine occurs most commonly in the form of grains, usually grouped into granular agglomerates. The loose grains of olivine often resemble tiny chippings of green glass. Crystals, which most often are columnar, are rare. Olivine turns yellow or red with oxidation. Mostly it occurs in substantial quantity in the upper layers of igneous rocks with a small part of silicon dioxide, especially in basalt. Sometimes olivines form independent rocks, called dunites. Dunite is the parent rock of diamonds in the Republic of South Africa and in Siberia. The olivine rocks in the Urals contain grains of platinum and chromite. Bohemian garnets also come from altered dunites. Small amounts of olivine are found in some meteorites, and it was also found present in basalt rocks in the Moon. About the year 1749 a meteorite hit the earth in the Yenisei region in Siberia. The local Tartar inhabitants considered the meteorite to be a messenger from heaven, and treated it with due respect. It was discovered in 1772 during an expedition led by the famous German natural scientist, Peter Pallas. Later it was definitely identified as meteoric iron, in which, to the astonishment of the scientists of those days, large chrysolites were also found. Some of the grains were actually good enough for grinding. The majority of the grains were covered by tiny crystal faces, but otherwise they were completely rounded and lacked any crystal edges. After this discovery, other iron meteorites with chrysolites were found in other regions of the earth; in honour of the discoverer of the first one, they were given the collective title of **pallasites**. The most magnificent specimens have come from localities in North America.

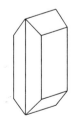

Frequently it is possible to see the alteration of olivine to talc, serpentine and to minerals with a content of iron, particularly limonite and haematite. Weathered olivines become brownish to brown from the start of the oxidation of the bivalent iron to trivalent. During a complete weathering process of olivine rocks, the insignificant original admixture of nickel becomes separated; during geological processes and lengthy geological eras important nickel deposits develop (such as the ones in New Caledonia).

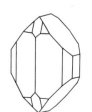

The main deposits of olivines are at: Zebirget, Arizona, Brazil, New Zealand, Norway, the Kozákov district in Bohemia, Forstberg near Mayen in the Eifel Mountains (West Germany). The transparent olivine varieties, which are also called **chrysolite** and **peridot** (an old French term of unknown origin) are still used as precious stones. Their popularity was never particularly constant. Olivine is softer than quartz and unsuitable for jewelry, in which it can be easily damaged. Its chief attraction lies in the beautiful yellow-green colouring and strong lustre, which grows stronger still with grinding. Olivine is ground into a tabular or graded shape and also as a brilliant and a rosette. It is inserted in orna-

244

mental pins, clips, ear-rings and pendants. Olivine can be effectively combined with other minerals, for example the Bohemian garnet, to make up a piece of jewelry. At the present time there is a revival of interest in chrysolite as a gemstone and its popularity is again growing.

A much rarer find than olivine are minerals of similar composition, **hortonolite** (a silicate of magnesium and iron), **forsterite** (magnesium silicate) and **fayalite** (iron silicate). Hortonolite, named after its discoverer, Horton, is basically olivine with a prevalence of iron over magnesium. It differs from the true olivine by its commonly brown colour, though it can be also yellowish, yellow-green to black. The main producers of this mineral are the O'Niel mine in the state of New York and several localities in the Republic of South Africa.

Forsterite, named after the American geologist Forster, is basically a magnesium variety of olivine and forms, together with olivine, isomorphous interstages. Most frequently it occurs in the form of minute crystals or grains. Generally it is colourless, whitish, yellowish to green. In the pure form it can be found in crystalline limestones and meteorites; interstages between forsterite and olivine occur in some dunites. The principal deposits of forsterite are in the lavas of Vesuvius, in Bolton, Massachussetts, USA, and in the Nikolai-Maximilian mine near Zlatoust in the Urals (USSR). Its occurrences near Raspenava in northern Bohemia are also interesting. There it is found partially altered to serpentines, called 'eozoon'. The forsterite crystals are arranged to form

278 Olivine — Podmoklice (Czechoslovakia); grains up to 2 cm.

greenish stripes, in which the alteration to serpentines takes place in irregular cracks, which resemble branches. For this reason it used to be taken for a fossil and was named accordingly.

Fayalite was named after the island of Fayal in the Azores, on whose shores it was discovered. Basically it is a ferrous variety of olivine. Most commonly it forms tabular crystals, coloured wine-yellow to olive-green. When weathered, it is reddish to brownish with a metallic lustre. It occurs in some granites. Fayalite also develops synthetically and in large quantity in the cinders of hot furnaces and in molten rocks. Its principal deposits are in the Yellowstone National Park, USA, and in the Mourne Mountains, Northern Ireland. It has been proved that on the island Fayal, where it was first found, fayalite was produced artificially.

279 Olivine, cut — Zebirget (Egypt); 30.8 × 18.8 mm; weight 66.43 carat.

280  Zircon — Jizerská meadow (Czechoslovakia); largest crystal 13 mm.

**Zircon** (280—281) has a unique position among precious stones. Its uses in the jewelry industry are not based on old tradition, as in the case of sapphire, ruby and spinel.

Zirconium silicate, $ZrSiO_4$, tetragonal. H. 7.5; Sp. gr. 3.9 to 4.8; commonly yellow-brown, brownish-yellow, yellow-green, reddish-brown to black, pale yellow to colourless; adamantine lustre, greasy on fracture. S. colourless.

Zircon was not fully utilized till the modern era and then mainly for its partial resemblance to a diamond. Until that time, pure, clear zircons were used for jewelry purposes only occasionally. They came onto the European markets under the name of 'Matara diamonds', named after the town of that name in Sri Lanka near which there are deposits of zircons. For a long time everyone assumed they were diamonds.

It seems strange that at first even the beautiful yellow-red, highly lustrous zircon variety — **hyacinth** — was not more popular in Europe; it has been mined from days of old in rich alluvials in Sri Lanka. Hyacinth was examined in greater detail at the end of the 18th century. The element zirconium was then discovered; however, proof that the clear 'Matara diamonds' from Sri Lanka were the same mineral, did not come till much later.

Zircon commonly occurs in tetragonal crystals, or in grains embedded in rocks. Small amounts (as a so-called accessory mineral) are frequently found in igneous rocks, especially in those rich in silicon dioxide and in alkaline metals, in potassium and sodium, for example. Zircon occurs in substantial quantities chiefly in nepheline-syenites and in pegmatites, in metamorphic rocks and in alluvials. Such secondary occurrences, especially in sandstones and gravels, are of the greatest practical use. Zircon is one of the most abundant heavy minerals, because of its mechanical resistance.

247

281 Zircon — Miass (Urals, USSR); crystal 4.5 cm.

The main producer of zircon is Travancore in India, where beach sands are recovered; zircon is their secondary, but just as important product. This deposit is, of course, only important for industrial uses. The most noted deposits of zircon of gemstone quality are, apart from Sri Lanka, Miass in the Urals (USSR), Renfrew in Canada, Henderson County in North Carolina (USA) and Cerro de Caldos in Brazil. The European localities are the Tirol, the Rhineland, Loch Garvee in Scotland, and southeast Norway, but European zircons are not of high enough quality to be considered gemstones.

Zircon is extracted for the production of zirconium, and sometimes of radioactive elements which may be present. Zirconium is used chiefly in the manufacture of heat-resistant materials, as a refractory, in abrasives, enamels, etc. Today it is used in nuclear reactors as a safeguard against corrosion by uranium. Clear zircon (hyacinth) and other coloured, but transparent varieties are used as gemstones. They are particularly valued for their lustre, which in the transparent zircons is noticeably high. This does not, however, apply to all the zircon varieties, for their unusual characteristic is a very wide fluctuation of their physical properties. The lustre is sometimes diamond but at other times is only vitreous.

**Andalusite** (282—283) was named after its first known occurrence in Andalusia (Spain). Before true andalusite was discovered, its variety — **chiastolite** (283) — (the name is derived from the Greek 'chiasmos' — arranged crosswise) was already known from the town of Santiago de Compostella, to which pilgrimages were often made. This variety was used for the manufacture of memorial objects, as a so-called 'stone of the Cross'. Chiastolite contains carbonaceous impurities arranged along rectangular axes so that when a transverse section is taken a coloured cross shows. Otherwise andalusite forms columnar or stalk-like crystals. Industrially important deposits are in Kazakhstan and in California. The other most noted deposits are in Austria, the Urals, Brazil and South Australia. Andalusite is used for the manufacture of heat-resistant substances and special porcelains. The variety from Brazil is used as a semiprecious gemstone.

Aluminium silicate, $Al_2SiO_5$, orthorhombic. H. 7.5; Sp. gr. 3.1–3.2; white, pink, and variously coloured; vitreous lustre; S. white.

282 Andalusite — Tirol (Austria); large crystal 7.5 cm.

283 Andalusite — chiastolite
Bimbowrie (Australia);
large crystal 4 × 3 cm.

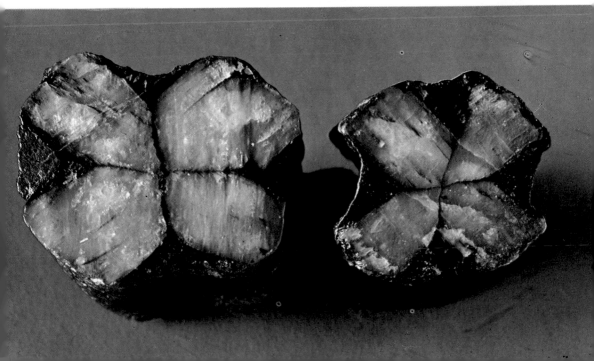

284 Disthene — kyanite — Bystřice (Czechoslovakia); 8 × 6 cm.

**Disthene** (284) received its name from the Greek 'di' — double, and 'sthenos' — strength, on account of its varied hardness in different directions. This is a very characteristic property of disthene. The hardness of tabular crystals is 4—5 along the prism but 7 across. The blue variety is called **kyanite**, the white to grey variety **rhaeticite**. The main occurrences are in metamorphic rocks formed under high pressure, especially in gneisses and pegmatites. It is commonly crystallized or in the form of acicular foliated aggregates. In many deposits it is found in association with staurolite or corundum.

Aluminium silicate, $Al_2SiO_5$, triclinic. H. 4—7 (varies with different directions); Sp. gr. 3.6—3.7; commonly blue (kyanite), white, also grey (rhaeticite), green, with vitreous lustre, pearly on some planes; S. white.

Chief deposits: Monte Campione near St Gotthard in the Alps, Bečov near Karlovy Vary in Bohemia, the Urals, Mount Beginse on the borders of Zaïre, the Sudan, Ouro Prêto in Brazil, the USA, India. Disthene is, like andalusite, resistant to high temperatures. It is therefore used in the manufacture of spark plugs and ceramics. Though a common mineral, its occurrences where extraction would be worth-while are rare (India and the Urals). Occasionally kyanite is used as a semiprecious stone (Switzerland).

**Sillimanite** has identical chemical composition, but is orthorhombic. It is commonly finely fibrous to acicular, and is coloured white. When mixed with quartz, it is called fibrolite.

**Topaz** (285—291) is one of the best known and one of the oldest gemstones, already popular in ancient times. In those days when only the Old World deposits were known, it was highly prized. Later, when rich deposits, yielding magnificent stones of substantial size were found in America, the interest in topaz fell considerably. Only a few of the coloured topaz varieties remain popular, mainly the honey-blonde stones, the pink and blue ones. This unfavourable change in the valuation of topaz was chiefly caused by the discovery of the largest ever topaz deposits in Brazil.

Aluminium fluor-silicate, $Al_2F_2SiO_4$, orthorhombic. H. 8; Sp. gr. 3.5–3.6; variably coloured, commonly yellowish to yellow, honey-yellow, blue, pinkish, pink to violet, colourless to clear, with vitreous lustre.

Topaz most probably received its name from the island of Topazos in the Red Sea (now called Zebirget) — see the entry for olivine. Some mineralogists believe the name was derived from the Sanskrit word 'tapas' — fire. The Romans gave the name 'topazus' to many ornamental stones of yellow colour (yellow olivines for instance). On the other hand, true yellow topaz was called 'chrysolithos' — golden stone. Only when the science of mineralogy became established, was topaz distinguished from chrysolite, the beautiful yellowish-green variety of olivine.

Topaz occurs most often completely colourless, white or grey. Slightly metalliferous admixtures are the cause of its varied colouring. The common yellowish hue, for instance, is caused by a small admixture of chromium. Bluish colour by traces of bivalent (ferrous) iron. A yellow-red colour is also typical, but the pink colouring of the Brazilian topaz is much rarer.

Topaz crystals are usually columnar, forming prismatic crystals made up of prisms with pyramidal endings. This termination is usually visible only at the end, for the crystals usually grow upwards. In other instances they terminate with a smooth basal surface, on which often the characteristic etchings originate. The prism faces are frequently vertically striated and display a prominent vitreous lustre. The larger crystals too are commonly transparent. Topazes from various deposits exhibit interesting surface irregularities and unevennesses, which occur through gradual dissolution. Topazes have a perfect cleavage. Many of the topaz crystals are surprisingly large. In the American Natural Science Museum in New York there is a perfectly bounded translucent crystal from Minas Gerais in Brazil, which weighs 300 kg and measures $80 \times 60 \times 60$ cm. In the museum of the Mineralogical Institute in Florence a pink crystal from the same deposit can be seen, weighing 150 kg. The topaz mines in Brazil boast of a find of a yellow crystal, broken into three pieces, which jointly weigh 140 kg. A giant blue crystal of topaz from Murzinka in the Urals is on show in the Mining College Museum in Leningrad.

The most famous cut and polished topaz is a stone called 'Braganza', which was first taken for a diamond. It weighs 1,680 carats and is set in the Portuguese crown. This topaz was found in 1740 at Ouro Prêto in Brazil. The largest cut topaz stones are in the collections of the Smithsonian Institution in Washington, and they all come from Brazil. The largest yellow stone weighs 7,725 carats, the blue 3,273 carats, and the yellow-green 1,469 carats. The most renowned topaz collection is among the treasures of Das Grüne Gewölbe in Dresden. It consists of many large topaz stones set in various pieces of jewelry, and some unset stones, and originally even uncut topazes were in this collection.

Topaz occurs even more frequently in granular or columnar form than in crystallized form. It is called **pycnite** (from the Greek 'puknos' — fat). Such varieties are commonly non-transparent, usually yellow, or slightly pink or green and therefore cannot be considered to be precious gemstones.

Topaz is a characteristic mineral found in cavities of some granites, especially the coarse-grained variety (pegmatites). It is also often found in cassiterite veins. In alluvials topaz occurs as a subsidiary mineral.

There are many famous deposits of topaz, some of which yield typically

285 Topaz — Schneckenstein (German Democratic Republic); large crystal 1 cm.

coloured varieties. The best known occurrences are in Brazil — and they have already been mentioned above. Topaz is found here mainly in the state of Minas Gerais, in two separate localities: the northeast part of the state, and in the vicinity of one of its largest cities, Ouro Prêto. Colourless topaz from the northeast part was first discovered from the gemstone placers as pebbles called 'pingos d'agoa' (drops of water); primary occurrences were later discovered in pegmatite veins in mica-schists and gneisses. The rich yellow and yellow-red topazes from the second locality were discovered in 1760 in weathered pegmatite veins which intruded into slate. Topaz is usually accompanied by clear quartz (rock crystal), smoky quartz, haematite, tourmaline and other minerals. Other occurrences are in the state of Rio Grande do Sul, Brazil, again in pegmatites.

Other famous deposits which yield topaz crystals which excel in size and beauty and are of varied colour, are in the USSR. They are often accompanied by beryl. The transparent, faintly blue crystals come chiefly from Murzinka in the Urals, near Sverdlovsk. They occur in association with smoky quartz, felspar and lepidolite in granite cavities. Miass, on the shore of Lake Ilmen in the Urals, is noted for a colourless topaz variety, the Adun-Chilon Mountains in Siberia for the yellow crystals, and Urulga (USSR) for the yellowy wine-coloured crystals. These are particularly large crystals. The European part of the USSR also has a number of new deposits where large crystals occur, such as Podoli in the Ukraine, and Volynj.

Of all the other world deposits the clear crystals of Japan are the most important, then the famous gemstone placers of Sri Lanka, which yield the clear, the yellow and the pale green topaz. Australia and North America also have topaz deposits at Romona (California), Ake's Peak and Nathrop (Colorado), Stoneham (Maine) and Utah, but they occur here in smaller quantities. The most noted European deposit of topaz is Schneckenstein in the Erzgebirge (East Germany) but it no longer has any practical significance. The occurrence of pycnite in the cassiterite veins of the Erzgebirge near Zinnwald in East Germany and Cínovec in Bohemia are of mineralogical interest.

Topaz's primary use is as a decorative gemstone. Colourless topaz is ground to a brilliant cut, whereas the coloured varieties are cut into an oblong shape.

It has been mentioned that the value of topaz is lower now than in the past, though these stones are still among the most popular ones. Larger

286 Topaz — Adun-Chilon (USSR); 9 × 6 cm.

pieces are not an exceptional rarity as with some other precious stones so generally, the value per carat of topazes does not rise with weight. But differences in colour often contribute to a topaz's value.

Topaz is appreciated for its hardness, transparency, pleasant colours and hues, and its purity as a mineral. The excellent cleavage enables the material to be easily divided, but makes the process of grinding all the harder. Some varieties tend to fade in sunlight, especially the richly coloured ones. This was known long ago, but not until much more recently was it proved that it is not the light, but the warmth of the sun's rays which cause the change in colours. From then on jewellers began changing the colour of topaz artificially by heating it. It is possible, for instance, to alter a Brazilian yellow or yellow-red topaz to a rose-red stone with a high refractive index and strong pleochroism.

In exceptional cases clear topaz is used as a raw material for the manufacture of special optical lenses. Topaz also has industrial uses as abrasive powder. The making of synthetic topaz is so far only of theoretical, and not of practical importance. This is why synthetic corundum, sometimes also spinel, is used to imitate topaz.

287 Topaz — Murzinka (Urals, USSR); 3 × 2 cm.

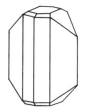

288 Topaz — Tonokamiyama (Japan); crystal 4 × 3 cm.

However, topaz is not just imitated, it is often used to imitate other precious stones of a much higher value. In both instances it is therefore necessary to keep topaz in mind whenever any precious stone is being identified. The difference between the colourless topaz and rock crystal, or leucosapphire, zircon or diamond can be determined fairly easily by measuring the refractive index. The refraction of light and therefore also the lustre of topaz, though higher than that in quartz, is substantially lower than in corundum, zircon and diamond. Phenakite is closely akin to topaz in optical properties and hardness. In colour topaz greatly resembles aquamarine, but differs from this mineral most conspicuously by its higher specific gravity. Colourwise it can also be mistaken for the blue zircon.

255

289 Topaz, cut — Brazil; largest 29.8 mm in diameter; weight 164.7 carat.

290 Topas, cut — Brazil; 43.9 × 30.6 mm; weight 197 carat.
291 Topaz, cut — Brazil; 66.9 × 59.4 mm; weight 1463 carat.

**Staurolite** (292—293) has a name derived from the Greek 'stauros' — cross, and 'lithos' — stone, for its crystals form twins which are intergrown in a cross-like fashion. This is why it was also called 'lapis crucifix', and why the twinned crystals were often worn as amulets. The most beautiful specimens came together with kyanite from the mica-schists of Monte Campione near Faido and from Lake Ritone in the canton of Ticino, Switzerland. They were sold under the name 'Baseler Taufstein' (Basle baptismal stone). They were fairly abundant, not particularly expensive and therefore were commonly worn.

Complex basic silicate of iron and aluminium, $Al_4FeOOH(SiO_4)_2$, orthorhombic. H. 7–7.5; Sp. gr. 3.7 to 3.8; brown, reddish-brown to black-brown, with a dull vitreous lustre; greasy on fracture; S. colourless.

It is interesting to note that individual columnar crystals of staurolite are rarer than the interpenetrated twins. They occur embedded in mica-schists and phyllites and in contact-metamorphic rocks of sedimentary origin with a surplus of aluminium oxide. Sometimes staurolite appears as a rock-forming mineral.

The main deposits, apart from Faido and Lake Ritone, are: Fannin County in Georgia and Lisbon and Franconia in New Hampshire, USA; Sanarka near Chkalov (formerly Orenburg) in the Urals (loose crystals in alluvials), Passeyr and Vipiteno in Tirol, St Radegund in Styria, Aschaffenburg in Bavaria (West Germany), Quimper in Brittany (France), Branná and Vysoký Jeseník in Moravia. Recently, large crystals of staurolite have been found in the Gorob mine in Namibia and also at Ducktown in Tennessee. The best examples of staurolite crystals, which also command the greatest attention of collectors, are at present found

292 Staurolite — Quimper (France); larger twin 4.5 × 4.5 cm.

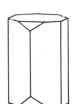

in the deposits in Georgia and New Hampshire. The individual crystals and the twins from Fannin County are of exceptional quality, but are relatively small. Only rarely do they reach 3 cm. Large, but rarely twinned crystals come from the deposits in New England and Windham (Cumberland County), Maine, which usually reach 5 cm in length. Magnificent twins occur also in Patrick County, Virginia. In Rio Arriba County, New Mexico, there are occurrences of parallel intergrowths of staurolite crystals according to two different laws.

Staurolite in which the iron is replaced by cobalt and magnesium is called **lusakite**. It is found in Lusaka, Zambia, where it occurs as a basic rock-forming mineral (it forms 30 per cent of the rock). There it is extracted as blue pigment.

293 Staurolite — Petrovice (Czechoslovakia); 6 × 4 cm.

**Titanite** (294) crystallizes in columnar, tabular or acicular form; it may also be massive, or can form granular aggregates. Titanite has some distinctive named varieties. **Sphene** is a transparent variety, which forms wedge-shaped crystals (the name in Greek means 'wedge'). **Castellite,** found in 1866 by Castelli (an Italian mineralogist), is a grape-yellow titanite. **Leucoxene** is an alteration product of ilmenite. The name is derived from the Greek 'leukos' — white, and 'xenos' — foreign, for it forms light stripes round the remains of the original or 'foreign' mineral (ilmenite).

Silicate of titanium and calcium, $CaTiSiO_5$, monoclinic. H. 5–5.5; Sp. gr. 3.4–3.6; commonly yellow, yellowish, greenish, brown, brownish-red to reddish, with a strong to adamantine lustre; S. colourless.

Titanite crystals, especially the sphene variety, most frequently occur in cracks of rocks, especially in crystalline schists, and are characteristic of the so-called 'Alpine paragenesis'. This is the classification for the family of minerals in veins and druses of cavities of metamorphic rocks, especially gneisses and amphibolites. Titanite occurs here in association with other minerals containing titanium, also with quartz, chlorites, felspars (mainly with albite or adularia), with amphiboles, epidote, prehnite, and with other silicates of aluminium, calcium, sodium and potassium. Titanite occurs as a secondary component of rocks in pegmatite veins, which are penetrating crystalline limestones, also in granodiorites, syenites and allied igneous rocks. Titanite however occurs in other rocks, such as in phonolite (castellite) and in greenstones.

The main deposits are in the vicinity of St Gotthard in the Alps (Switzerland), near Achmatovsk in the Urals and on the Kola Peninsula in the USSR. Flat brown crystals and grains of titanite are found here in the Khibiny massif in syenites with a poor content of silicon dioxide, which are often named according to the locality of their occurrence. An example is 'luyavrites', named by a famous Norwegian mineralogist and geologist W. C. Broegger (born in 1831) after the place of discovery, the Khibiny tundra Luyavrurt, which contains as much as 16 per cent titanite. The best titanite crystals occur in the local rock — coarse-grained nepheline syenites named chibinites by the British geologist and petrographer, Sir A. G. Ramsay (1814—1891). Titanites occur even in rischorites and foyaites of Khibiny — and they are some of the loveliest in the world. Usually they are accompanied by their rare associates, such as **murmanite** (hydrated silicate of titanium, manganese and sodium), **lomonosovite** (silicate and phosphate of sodium, titanium and manganese) and **fersmanite** (a complex basic silicate of titanium, calcium and sodium). In the Khibiny tundras on the Kola Peninsula, titanite is recovered as a by-product during the mining of apatite. Other noted deposits of titanite are near Kragerö in Norway, in the Plauen valley near Dresden, East Germany, near Čáslav in Bohemia, in the Tilly Foster iron mine in the state of New York (USA), and in several other places.

Titanite, when found in large enough quantity, is important primarily as a titanium ore (it contains up to 24.5 per cent titanium). Titanium is used in the production of ferrotitanium alloys, which are very resilient, especially against acids. Compounds of titanium are used as yellow and yellow-red glazes in pottery manufacture, also in the manufacture of synthetic fibres and in the textile industry. The oxide is the basis of 'titanium white' (a brilliant white pigment). Titanium is exceptional for its low weight, its resistance to corrosion and for its hardness. In some deposits titanite is also a valuable source of rare earths. These are usually part of titanite varieties such as yttrotitanite or keilhautite, grothite, alshedite and eukolite-titanite.

The attractively coloured titanites are sometimes cut and polished, usually in the brilliant cut; they then serve as not particularly valuable, but fairly rare gemstones. Often they are used as imitation topaz stones, gold beryl, olivine, chrysoberyl, vesuvianite and garnet (demantoid), for instance.

Titanates are very similar to titanite, but according to the modern

294 Titanite — Ofenhorn (Switzerland); crystals up to 10 mm.

structural system, they are included with binary oxides. **Ilmenite** is the most important member of this class; the name comes from the Ilmen Mountains, USSR. It is a rhombohedral iron titanium oxide, $FeTiO_3$, which crystallizes most commonly in thick-tabular, lamellar or foliated form, like the crystals of haematite. In other instances ilmenite occurs ingrown in rock as irregularly bounded grains, or in fluvial deposits, particularly in sands, as iserine for instance (from the German name of the Bohemian river Jizera — Iser). There it developed mostly from altered rutile. Occasionally it is intergrown with magnetite ('titano-magnetite') or with haematite. It is black, with a sub-metallic to metallic lustre, and occurs in basic magmatic rocks, pegmatites, and ore-veins. Chief producers: the Urals, Egersund in Norway and Canada.

**Hemimorphite** (295) has been mined from days of old from the upper parts of zinc- and lead-based ores, chiefly associated with smithsonite. It was often assumed to be the same mineral and both were classed under the same name of calamine. In the second half of the 18th century it was discovered that there were two different minerals under the heading of calamine — a zinc carbonate and a zinc silicate, which often closely resembled each other.

Hydrous basic zinc silicate, $Zn_4(OH)_2Si_2O_7 \cdot H_2O$, ortho-rhombic. H. 5; Sp. gr. 3.3–3.5; commonly colourless to clear, or yellow to brown; strong vitreous lustre; S. colourless.

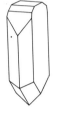

The silicate was the rarer of the two, and was named hemimorphite on account of the hemimorph development of its crystals. This unusual form, which is typical of only a few minerals, means that the crystals are terminated by dissimilar faces. Hemimorphite most commonly forms crystalline crusts and layers, also massive, granular, rounded and reniform aggregates, concentrically striated, or finely needle-shaped, fibrous or stalactitic, and rarely fan-shaped clusters of crystals.

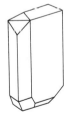

Hemimorphite most frequently occurs as the product of the oxidation of the upper parts of sphalerite, accompanied by other secondary minerals which form the so-called 'iron cap' or 'gossan'. The origin is by the process of metasomatism, that is by the gradual replacement of the easily soluble limestone with less soluble matters brought by circulating waters.

The regions on the Belgian-German border are well known for their deposits of hemimorphite of metasomatic origin, especially Vieille Montagne in Belgium and Aachen in West Germany. Other deposits are near Tarnovice in upper Silesia (Poland), near Phoenixville (Pennsylvania), Elkhorn (Montana), Leadville (Colorado) and Organ Mountains (New Mexico) in the USA, and in several localities in North Africa. Further hemimorphite occurrences are in Nerchinsk in Siberia, Rabelj in Slovenia (Yugoslavia), Bleiberg in Carinthia (Austria), Matlock in Derbyshire (England), etc. Hemimorphite is an important ore of zinc and contains up to 54.2 per cent of the metal.

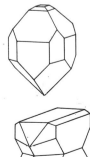

295 Hemimorphite — Nerchinsk (USSR); actual size of detail 7 × 6 cm.

**Vesuvianite** (296—297) could be called the mineral of many names. The main name was given from its locality, Mount Vesuvius in Italy, where it is found in volcanic rocks formed by eruptions. Vesuvianite's other name — **idocrase** — is derived from the Greek 'eidos', which means likeness, and 'krasis' — composition, for this mineral has a similar composition to several garnets, and is often mistaken for them. A number of other names are based on the various localities of its deposits. It is known as **egeran** after Eger — the German name for Cheb in Bohemia where it was discovered. The nearby deposits at Hazlov are world famous. Here it occurs in association with the hessonite variety of garnet and with wollastonite and with albite. As the variety found here is always brown and coarsely acicular, the name egeran is sometimes applied to all brown or acicular varieties of vesuvianite. The poet, Goethe, who was also a keen collector of minerals, was so taken with the beauty of vesuvianite, that he composed a long poem about it. Another name — **vilyuyite** — comes from the mineral's occurrence on the banks of the river Vilyuy in eastern Siberia, and the name **californite** from California. For quite a while vesuvianite has been mistaken for tourmaline because of its external characteristics, and also for olivine, topaz or garnet. The eventual recognition of vesuvianite as a separate mineral was by Abraham Gottlob Werner, a mineralogist from Freiberg, and he named it vesuvianite. At the same time vesuvianite was identified as a highly complex silicate. Its chemical formula can only give an approximate composition, for this mineral contains often a number of admixtures of other elements, particularly manganese, chromium and beryllium.

Vesuvianite's crystals are usually prismatic and columnar, or elongated needle-shaped, parallel or in different directions; it also forms granular or compact aggregates, and occasionally pyramidal crystals. The crystals are rather fragile, and usually only translucent or non-transparent. Different varieties are variously coloured; for instance egeran is brown, californite green, vilyuyite black-green. The yellow variety was named **xantite**, the pale blue variety **cyprine**. A typical mineral of limestone and dolomite in contact with igneous rocks, it occurs there in association with other contact minerals, mainly silicates of calcium. Less frequently it is found in cracks of crystalline schists.

From the many localities of vesuvianite's occurrence, the following are of importance, apart from those already mentioned: Göpfersgrün in West Germany, Arendal in Norway, Monzoni near Predazzo in northern Italy, Cziklov in Romania, Bludov in Moravia, the Achmatov mine in the Urals, Brazil and Kenya. The last two supply chiefly the gemstone varieties. As a precious stone it is rather rare and little used. But sometimes attractively coloured pieces of contact rock enclosing vesuvianite are cut and polished. Californites are treated similarly and made into ornamental stones of lower value for jewelry purposes, or they are used in the manufacture of smaller artistic articles.

The deposits of californite, which occur in several localities in California, are much sought after by collectors. The most productive deposits are in Siskiyou County. There are many other important deposits of vesuvianite in the USA, a number of which have only been discovered recently. The interest of mineral collectors is centred on every one of them. Notable deposits occur at Magnet Cove (Garland and Hot Spring Counties) in Arkansas, where imperfect crystals bipyramidally developed occur in gigantic sizes up to 30 cm long. Perfectly formed crystals are found on contacts of the pale blue limestones at Helena, Lewis and Clark County, Montana. They do not exceed 4 cm in length. Similar crystals, but smaller still are found again in blue limestones in quarries at Riverside, California. Dark brownish-green columnar crystals with a perfect square cross-section have been discovered at Sanford (York County) in the state of Maine. They are up to 3 cm in length and

Highly complex basic silicate of calcium, magnesium and aluminium, with iron, tetragonal. H. 6.5; Sp. gr. 3.27–3.45; commonly green-brown, brown or green, variably shaded, with vitreous lustre, greasy on fracture; S. colourless.

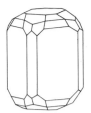

form magnificent druses in limestones. It is possible to find crystals of exceptional beauty in the asbestos-bearing rocks by Eden Mills (Lamoille County) in Vermont, and large, imperfect crystals in marble near Olmstedville (Essex County) in the state of New York. The cyprine variety has also been extracted from the noted deposit at Franklin (Sussex County) in New Jersey.

Collectors also frequent deposits in Mexico and Canada. Imperfect columnar crystals up to 8 cm in length and of a yellow colour are found with the grossular variety of garnet in the well-known deposit at Xalostoc, Morales and by Lake Jaco in the Chihuahua province of Mexico. The occurrences of the yellow-brown vesuvianite are of gem quality; it is found in coarse columnar crystals in Quebec, Canada.

Collectors still hold hope for the classic vesuvianite deposits in Europe. It is possible to find glassy olive-green columnar crystals of a perfect square cross-section and up to 3—5 cm in length in Norway, particularly

296 Vesuvianite — egeran — Hazlov (Czechoslovakia); 7 × 6 cm.

in the vicinity of Kristiansand and at Eiker near Oslo. The new discoveries in northern Italy are of a similar type; beautiful crystals are found chiefly by Canzocoli near Predazzo and near Ala in Piedmont. Crystals from the latter deposit are noted for their beautiful green colour and transparency. The crater Monte Somma on Vesuvius also yields beautiful crystals up to 2 cm. They are commonly terminated in pyramidal form. Beautiful green to blue-green coarsely columnar crystals are still being found near Achmatovsk in the Urals. Such specimens are valued even more by collectors, for their parent rock is a magnificent, delicate green, so the whole effect is quite unusual.

297 Vesuvianite — Egg (Norway); crystals 15 mm.

**Epidote** (298—299) was given the name from the Greek 'epidosis' — an addition, because it had been for a long time mistaken for tourmaline and its subsequent determination meant a new addition to the mineral system. The French mineralogist and crystallographer René Just Haüy was the first to distinguish them in 1901. The second name it is known under — **pistacite**, is based on its green colour, so like the colour of pistachio nuts.

Highly complex basic aluminosilicate of calcium and iron, monoclinic. H. 6–7; Sp. gr. 3.3 to 3.5; commonly greenish, pale green, brownish-green to black, paler with smaller admixture of iron; vitreous lustre; S. grey.

Epidote forms columnar twinned crystals, needle-shaped granular and compact aggregates, or crusts and coverings. It is also a contact mineral, but occurs in other metamorphic rocks as well, which have originated

265

299  Epidote — Knappenwand (Austria); largest crystal 8 cm long.

under low temperatures but raised pressures (epidotes with a small iron content), especially in schists and skarns. Here it is fairly abundant. Sometimes epidote forms independent rocks, which are called epidosites. When crystallized it is green, but massive varieties are yellow-green to yellow. Epidosites form nests and fillings in amphibolites.

The main deposits are in: Knappenwand in Sulzbachtal of Salzburg, Sobotín in Moravia, Kowary in Polish Silesia, also Achmatovsk in the USSR. Epidote is also found on Prince of Wales Island, Alaska. As it lacks hardness and distinct colour, it is little used as a precious stone. As a rule only the dark green epidotes are cut and polished, for instance those from Salzburg and Sri Lanka. It has no other practical significance. Some beautifully crystallized specimens, chiefly from Knappenwand, are greatly desired by collectors.

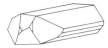
Epidote which contains the rare-earth elements caesium and lanthanum is called **orthite** (from the Greek 'orthos' — straight, on account of the shape of the crystals). It occurs in the form of black or brownish grains with a resinous lustre, embedded in granites and pegmatites.

266

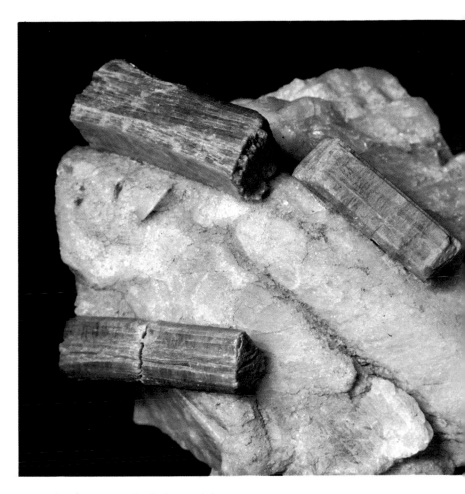

300 Zoisite — Zermatt (Switzerland); crystals 4 cm.

**Zoisite** (300) aroused no great interest as a mineral until the year 1967. It was then that an Indian tailor, Manuel d'Souza, found by chance a beautifully coloured blue gemstone near his native town of Arusha, southwest of Kilimanjaro. This discovery triggered off exceptional interest, and many newspapers and magazines blew this up as sensational news, calling the mineral 'the blue African treasure'. It was named **tanzanite**, after the country of its discovery, by the noted jewellers Tiffany of New York, keenly interested in the beautiful stone.

Basic silicate of calcium and aluminium, $Ca_2Al_3OH(SiO_4)_3$, orthorhombic. H. 6; Sp. gr. 3.2 to 3.4; grey-white, greenish, brownish, reddish (thulite); rarely blue (tanzanite); vitreous lustre, pearly on cleavage planes; S. colourless.

The numerous articles, which appeared in newspapers and magazines all over the world, were often accompanied by magnificent coloured illustrations; it was obvious that this was a comparatively rich occurrence of some magnificent and large crystals, closely resembling sapphires. But for a long time the actual identity of the mineral in question was not clear. Tanzanite was thought to be first cordierite, then dumortierite (a complex borosilicate of aluminium).

Members of the German society for precious stones in Idar-Oberstein in Germany, where the mineral was taken for cutting and polishing, at last gave the stone its exact identity. They discovered that tanzanite was a variety of zoisite — a normally rather inconspicuous mineral of usually

267

greyish or brownish colouring. The blue colour of tanzanite is obviously caused by a large admixture of strontium. The mineralogical institute in Heidelberg made a detailed study of this mineral and identified and determined all the chemical and physical properties. Apart from the high content of strontium, tanzanite was found to have a low content of iron, and a number of new findings about the chemical composition were discovered. It was unambiguously ascertained at the same time that there is no reason to consider tanzanite as a new mineral in the true sense of the word, although it is a most interesting stone. The blue variety of zoisite was, until then, quite unknown. It is therefore no exaggeration when tanzanite is mentioned as the youngest gemstone.

Zoisite (named after a Slovenian collector, Baron Sigismund Zois von Edelstein, who financed mineral-collecting expeditions, during one of which zoisite was discovered), commonly forms massive, striated, coarsely and finely elongated to fibrous aggregates. The crystal form is rare. The pink variety, coloured by the presence of manganese, is called **thulite** (after Thulé, the ancient name for lands in the far North, including Norway, where thulite is found). Zoisite occurs in the cracks of deep igneous rocks, where it forms as a secondary mineral through alteration of the original minerals of which the rock is composed, and in contact limestones.

The main occurrences are Zermatt in Switzerland, Norway, Gefries in West Germany, Borovina in Moravia, Saualpe in Carinthia (Austria), Lake Gravee (Scotland) and California, Ducktown (Tennessee) and Goshen (Massachusetts) in the United States.

The pink Norwegian thulite has been used for quite some time for jewelry purposes, and the Californian grey-green zoisite has also come into its own — but otherwise they have no wider use. The stones are usually cut into the cabochon shape, and are sometimes engraved. Tanzanite is a better gemstone than zoisite. It is popular chiefly for its blue colouring, whose richness is further accentuated by firing the stone. Though the refractive index is lower in tanzanite than in normal zoisite, it is still rather high, similar to corundum, which raises its gemstone quality, just as the low double refraction does. But the low hardness and obvious cleavage are a disadvantage.

Tanzanite has climbed into the class of exceptionally precious stones and is valued as such. In 1969 the price of the ground stones in Idar-Oberstein fluctuated between 40 and 50 DM per carat for the light coloured stones, and up to 800 DM for the dark varieties, according to their quality and size. Its value has not fallen even in current times, when rumours are circulating that the deposits are almost exhausted. The largest tanzanite found to date weighs 126 carats, and its value is estimated at 100,000 DM. It is said that Elizabeth Taylor, the American film star, owns the most beautiful tanzanite piece of jewelry. It is a necklace with five tanzanite stones.

**Axinite** (301) was first identified in 1797 when found in the Alps by R. J. Haüy. Axinite crystals are formed with sharp, acute wedge-shaped edges, rather resembling an axe, and this has provided the mineral with its name. Axinite has an exceptionally high vitreous lustre, and this characteristic led to yet another name 'vitreous schorl' for it was originally considered to be a variety of tourmaline, similar to schorl.

Highly complex basic borosilicate of calcium and aluminium, with iron, magnesium and manganese, triclinic. H. 6.5 to 7; Sp. gr. 3.3; commonly blue-brown, blue, grey, yellow-green, with strong vitreous lustre; S. colourless.

Most commonly, axinite occurs as a metasomatic mineral, on contacts, in skarns, basic rocks and in veins. Occasionally it is also found in greenstone and pegmatite. The main deposits are at Le Bourg d'Oisans in the French Alps, the Harz Mountains, Zbraslav near Prague (Czechoslovakia), St Just (Cornwall), and Franklin (New Jersey) and Luning (Nevada) in the USA. It is used as a semiprecious stone.

**Datolite** resembles axinite in several ways; it is a basic silicate of calcium and boron $CaBOHSiO_4$, which forms monoclinic colourless crystals of varied shapes or completely compact masses. In crystallized form it occurs chiefly in cracks of rocks poor in silicon dioxide; the massive form is rare and occurs as contact rock. The main deposits are at Andreasberg, East Germany and New Jersey, USA.

301 Axinite — Le Bourg d'Oisans (France); crystals up to 3 cm.

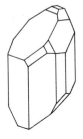

302 Benitoite — San Benito County (California, USA); crystals 14 mm.

**Benitoite** (302) is named after its one and only proper deposit near the San Benito Mountain in California's Diablo Range. For a long time it was taken for a sapphire. In 1907, however, it was identified by the American mineralogist Landerback as a new mineral. It did not take long before it became highly prized as a gemstone. This is understandable, for it has an outstandingly high lustre, higher than sapphire, and exquisite blue colour, for which it was also aptly called 'the heavenly stone'. But cutting and polishing presented difficulties. Benitoite crystals happen to show strong pleochroism, which means that they have different colours in different directions. They are richly blue in the direction of the main crystal axis, but in the vertical direction to the axis they are almost colourless. Benitoite occurs extremely rarely in cracks of crystalline schists. In San Benito, California, it is found in altered serpentinites, which are chiefly composed of sodiummagnesium amphibole crossite, and impregnated with crystalline natrolite. Benitoite is found here on walls of fissures, in association with rare minerals, such as the black neptunite (silicate of titanium, iron and sodium) and the tiny orange-brown crystals of jaaquinite (complex silicate of sodium, calcium, titanium and iron). Its practical use is rather limited as its occurrence is so infrequent. The value of benitoite seems substantially high, especially as its crystals are not commonly large. The largest crystal found up to date weighed only 7 carats.

Barium titanium silicate, $BaTi(SiO_3)_3$, rhombohedral. H. 6.5; Sp. gr. 3.7; blue, often with violet shades, less commonly colourless, with vitreous lustre; S. colourless.

270

303 Dioptase — Altyn Tybe (USSR); crystals up to 3 cm.

**Dioptase** (303) was first described by the French mineralogist and crystallographer, René Just Haüy in 1797. Long before that time it was known and admired for its vivid emerald-like colouring, caused by the presence of copper. It could perhaps be occasionally mistaken for an emerald, but its negligible hardness (5) gives its identity away.

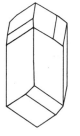

Hydrous silicate of copper, $Cu_3(SiO_3)_6 \cdot 6H_2O$, rhombohedral. H. 5; Sp. gr. 3.3; green, with vitreous lustre; S. green.

Dioptase most frequently forms columnar crystals of perfect cleavage in copper-ore deposits. The main occurrences are in Shaba (Zaïre), Altyn Tybe in Kazakhstan in Soviet central Asia, Băita (Romania), Copiapó (Chile) and Otavi (Namibia). Dioptase is not widely used, but is a highly valued precious stone; the transparent variety shows an outstanding dispersion of rays of light and consequent wonderful play of colours (known as fire). The low hardness is a disadvantage.

**Catapleite**, a hexagonal hydrous silicate of zirconium and sodium $Na_2ZrSi_3O_9 \cdot 2H_2O$, resembles dioptase and is rarer still. Catapleites in which sodium is replaced by calcium are more common. This mineral forms tabular crystals of perfect cleavage, which are pale yellow to yellow-brown. It is found in pegmatites of Norway.

271

**Beryl** (304—311) and its exquisitely colourful varieties, especially the emerald, are among the longest used and most widely used stones. It is said that the beautiful Cleopatra had an emerald engraved with her own picture. Egyptian emerald mines were being worked during her lifetime (69—30 BC); but long before this they were mined by the ancient Egyptians (c. 1650 BC). The ancient Romans searched for emeralds in Egypt later, then the Arabs and, later still, the Turks.

Beryllium-aluminium silicate, $Al_2Be_3(SiO_3)_6$, hexagonal. H. 7.5—8; Sp. gr. 2.63–2.80; commonly green or differently coloured, for instance yellow to greenish-yellow (heliodor), blue-green (aquamarine), pink (morganite), vivid green (emerald), occasionally colourless, with vitreous lustre; S. white.

It is believed that the Peruvians worshipped an emerald as big as an ostrich's egg. The Indians recovered emeralds in Colombia long before the appearance of the Spaniards, who took possession of the mines in 1537. In the Natural History Museum of Vienna, a beautiful druse of emerald crystals can be seen, which measures $18 \times 13$ cm. It is said they belonged to Montezuma, who was the Aztec emperor from 1502 to 1520, and that they found their way to Vienna from the Prague collections of Rudolf II, who was the German emperor from 1576 to 1612.

The druse found in the treasures of 'Das Grüne Gewölbe' in Dresden comes from the same source. In the sanctuary of the Buddhist temple in Kandy, Sri Lanka, there is an ancient statuette of Buddha carved from a single piece of the gem.

**Emerald** (305) is a richly green variety of beryl; the clear, light blue-green variety is **aquamarine** (306—307; 311); the yellow, perfectly transparent, golden beryl is called **heliodor**; the pink variety is **morganite** (308—309), which owes its colour to the admixture of a rare element — caesium. Common beryl is of an inferior quality, not suitable for gem purposes. Most commonly beryl forms simple, prismatic hexagons, often elongated and columnar, more rarely thickly tabular. Sometimes the crystals reach several metres in length. In Maine (USA),

304 Beryl — Maršíkov (Czechoslovakia); $5 \times 3$ cm.

305 Beryl — emerald — Tokovaya river (USSR); $5 \times 5$ cm.

306 Beryl — aquamarine —
Sta. Rita (Brazil); largest
crystal 40 mm long.

307 Beryl — aquamarine — Adun-Chilon (USSR); crystal 9 cm long.

308 Beryl — morganite — Pala (California, USA); 10 cm.

crystals were found measuring up to six metres and weighing as much as 1.5 tonnes.

Beryl most commonly occurs embedded in coarse-grained granites, called pegmatites, or in rocks of similar origin. The best examples of emeralds come from calcite veins and mica-schists; the most beautiful aquamarines come from some of the rock debris and alluvial deposits, which are formed through the weathering of the parent rock, the coarse-grained granite.

The most noted emerald deposits were in the region south of Koseir in Egypt (1). Today these mines are deserted. The most beautiful specimens of emeralds are found now in the vicinity of Musso, Colombia, South America, where the largest currently worked deposit of this stone exists. The Urals also boast of extensive deposits, and the biggest emerald crystals come from the Transvaal. They measure up to 25 centimetres.

Magnificent aquamarine examples come from Brazil, from Nerchinsk in Siberia and from pegmatites and granites near Murzinka in the Urals. The rich south Brazilian deposits of precious stones yield fairly frequently substantially large aquamarine crystals of gem quality. They are a magnificent pale blue. In 1910 a green-blue, perfectly clear aquamarine was found in the local pegmatites; it was nearly half a metre long and weighed over 110 kg. It was sold for 25,000 dollars. The **maxaxite** variety is particularly highly prized; these deep blue aquamarines have been thus coloured by the admixture of boron. They are called after Maxax in the state of Minas Gerais, where they were discovered. The Minas Gerais deposits are the most interesting occurrences of aquamarine from the mineralogical and practical view, for all the other coloured varieties of beryl of gem quality are found with it.

The largest ground aquamarine is in the Smithsonian Institution in Washington (1,000 carats). The National Museum in Prague has a pale

309 Beryl — morganite, cut — California (USA); 27.2 × 21.4 mm; weight 53.42 carat.

310 Beryls, green, cut — actual size of largest 51.8 × 41.8 mm; weight 421.6 carat.

blue aquamarine, weighing 991 carats in its collection. A green aquamarine in the British Museum in London weighs 875 carats. All these large stones originate from Brazil. The largest existing private collection of big aquamarines consists of 30 stones over 100 carats. In this collection there are also beautiful specimens of other coloured beryl varieties.

The value of aquamarine stones is nowhere near as high as the value of emeralds. This applies also to the differently coloured gem varieties of beryl, though these are stones of exceptional beauty and have been much sought after, particularly in recent years. There is one exception in this evaluation, and that is the most rare, richly grass-green beryl, whose colour differs from the colour of both emerald and blue-green aquamarine. This particular stone is one of the most expensive gemstones and its value nears the value of an emerald. The most important deposits of the coloured varieties of beryl are at Pala in California and Marahitra in Madagascar, from where the unique pink beryl (morganite) comes. Crystals of the variedly coloured and generally almost non-transparent beryl are frequently considerably large. The biggest known crystal comes from the State of Maine, USA; it weighs 18 tonnes. The largest ground greenish-yellow beryl (heliodor or golden beryl), which was found in Brazil, is now in the collection of the Smithsonian Institution. This stone weighs 2,054 carats. The biggest morganite from Madagascar weighs 598 carats and is in the British Museum in London.

Beryl is not only a prized gemstone, but also virtually the only raw material for the manufacture of one of the lightest metals — beryllium, which is used in the production of alloys. Pure beryl contains up to 14 per cent beryllium.

311 Beryl — aquamarines, cut — Brazil; largest 68.5 × 55.8 mm; weight 990.6 carat.

**Cordierite** (312) was defined as a distinct mineral in 1813 and named after Louis Cordier, the French mineralogist, geologist and mountaineer, who was the first to describe it with accuracy in 1809. Long before this, cordierite was known and sought after as a gemstone, chiefly in Sri Lanka, where it was considered to be a sapphire variety. The lighter shades of cordierite were sold in Sri Lanka under the title of 'water sapphires'. From the number of other names bestowed upon this mineral, it is worth mentioning dichroite (dual-coloured), because it is distinctly pleochroic.

Aluminosilicate of magnesium, $Mg_2Al_4Si_5O_{18}$, orthorhombic. H. 7–7.5; Sp. gr. 2.6; usually brownish-blue, yellowish, brownish-green, grey-blue, blue to violet, strongly pleochroic, sometimes even to the naked eye; also brown, with greasy, vitreous lustre; S. colourless.

Cordierite occurs in the form of grains, short-columnar crystals, or in massive form. It is comparatively common, especially in gneisses of sedimentary origin (para-gneisses). With the effects of weathering it changes into various mica varieties. The main deposits are in Bodenmais in Bavaria, the surroundings of Murzinka in the Urals, Haddam and Guilford in Connecticut, USA, Japan, and Sri Lanka. Cordierite crystals from Bodenmais are commonly partially altered and reach 4 cm in length. Crystals found in Orijarvi in Finland have not gone through the process of transformation and are up to 5 cm long. At Kragerö in Norway, cordierite forms impressive vitreous masses of substantial dimensions. They are commonly permeated with inclusions of red haematite. The best gemstone material today comes from Madagascar.

Cordierite is not often used as a gemstone. It is ground to a brilliant lenticular cut, so that full advantage is taken of the stone's pleochroism. The stones which are violet in one direction and grey in another, are a most interesting curiosity.

312 Cordierite — Bodenmais (Federal Republic of Germany); 8 × 8 cm.

313 Sekaninaite — Dolní Bory (Czechoslovakia); 4 × 3.5 cm.

**Sekaninaite** (313) was known from the past to occur only in one locality — in Dolní Bory in

Ferrous aluminosilicate, $Fe_2Al_4Si_5O_{18}$, orthorhombic, H. 7; Sp. gr. 2.77; violet to grey-blue, with vitreous lustre; S. colourless.

Moravia. Until 1968 it was taken for cordierite. Only a short while ago it was named after a Czech mineralogist, Josef Sekanina. Next to brazilianite and bukovskyite, sekaninaite is probably the only other newly discovered mineral, which appears in fairly substantial occurrences and therefore has great interest for collectors.

It appears in the form of imperfect, large crystals and in coarsely granular or compact aggregates of a violet or grey-blue colour, which tend to turn green or yellow through the effects of weathering. Like cordierite, sekaninaite too alters into different types of mica, chiefly into muscovite. Some forms used to be described misleadingly as pinite (after the deposit of cordierite pseudomorphs from the Pini mine near Schneeberg in East Germany), or as gigantolite (because it forms pseudomorphs after the gigantic crystals of cordierite). Similar pseudomorphs occur in several localities in Czechoslovakia, particularly near Jihlava and by Dyleň in the Bohemian Forest. But they are altered to such a degree that it is impossible to determine whether sekaninaite or cordierite was the original mineral.

Pegmatite in contact with gneisses of sedimentary origin (paragneisses) are the parent rock of sekaninaite. Sekaninaite has also recently been discovered in Japan.

**Tourmaline** (314—318) is the latest European arrival of all precious stones which are in current use. It was not until the beginning of the 18th century that it was transported from Sri Lanka by Dutchmen. Its name comes from the Sinhalese 'toramalli'. Tourmaline has a very strange characteristic that was incomprehensible in those days. When heated, it attracts ash from the fire. Today, it is known that this is due to electricity, which originates in tourmaline on account of its exceptional internal structure. The columns of tourmaline are charged, when heated, positively at one end, and negatively at the other. This is called the pyroelectric capability of tourmaline.

Highly complex basic borosilicate of a number of elements, especially sodium, aluminium, magnesium, calcium and iron with a changeable composition, rhombohedral. H. 7–7.5; Sp. gr. 3–3.25; colours see text; vitreous lustre, resinous in fracture; S. colourless.

The Sri Lanka variety is reddish to vividly red (**rubellite**) (314), but the less conspicuous black tourmaline (**schorl**) (316, 317) was known long before the Sri Lanka variety. European miners and stoneworkers were familiar with schorl in many localities, but its dull colouring did not arouse their curiosity. No one ever dreamt that the pink 'toramalli' of Sri Lanka and the black schorl were an identical mineral.

The diversity of colour made tourmaline popular as a precious stone

314 Tourmaline — rubellite — Pala (California, USA); 10 cm.

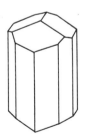

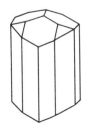

278

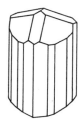

315 Tourmaline — dravite — Dobrova (Yugoslavia); crystals 6 mm.

in ancient times. It is possible to judge from excavations and written descriptions that it was favoured highly in the past. But it was so often mistaken for other minerals, that gradually it fell into oblivion and by the Middle Ages it was again quite unknown in Europe.

From the chemical point of view, tourmaline is one of the most complex minerals and its composition varies, for its different components intermingle and replace each other. This characteristic, of course, makes tourmaline far more varied in colour than other minerals. Schorl is the most common tourmaline variety, whereas the colourless **achroite** is the rarest one. The most popular pink to scarlet rubellite is often intensively coloured, and resembles a ruby. Rubellite is chemically distinguishable by the presence of lithium. The green **verdelite's** richness of colour often varies, as does the blue **indicolite's**. The brown **dravite** is rather scarce (315). Even in a single crystal there may be striking banding of colours. The columnar crystals often display alternate transverse coloured bands, or alternately, these are visible on a cross-section of the columnar crystal. The core, therefore, is commonly coloured differently to the exterior zones. Tourmaline can be rightfully considered one of the gayest gemstones. The colour of tourmaline crystals can vary in different directions. This is caused by tourmaline's tendency to absorb light in the direction perpendicular to the vertical axis, but never in the direction of the vertical axis. Such a characteristic can be best observed in a sphere

279

ground from a tourmaline crystal. This property is called pleochroism, or, to be more precise, dichroism, for in this case only two colours are concerned. Such a selective absorption of light can be observed in translucent and transparent crystals; the black-coloured tourmalines display this characteristic only in thin sections. Dichroism is of course shown by other minerals too, as long as they are coloured and anisotropic (ruby, for example); of all the gemstones, tourmaline, however, displays the most intensive dichroism, which in many instances can be seen with the naked eye. It is most necessary that all grinders are well acquainted with dichroism, for in their work they must select the correct position to ensure the tourmalines whose colour is too deep will become more transparent and those whose colour is too light, deeper.

Tourmaline crystals are usually long-columnar, more rarely short-columnar, and are often deeply striated. More frequently still, the crystals are needle-shaped.

Tourmaline is a very common mineral of granites and pegmatites, and appears secondarily in fluvial deposits. Mostly it is accompanied by lithium-mica (lepidolite) with its pink colouring, by beryl, topaz and apatite. At times it forms tourmaline rocks of a finely needle-shaped to granular composition, usually when closely associated with zinc veins.

The richest deposits of the gem varieties of tourmaline are in South American pegmatites. Perfectly formed rubellite columnar crystals were known especially from the district of Pala in California. Today this deposit is almost exhausted. The most vividly colourful crystals of tourmaline are found on the Italian island of Elba. Alternating bands of colour often make these crystals unusually attractive and gay. On Elba, crystals which have black ends, but a green and pink core, are a common occurrence and are called 'Moor's heads'.

Brazil is the home of the green variety, which occurs chiefly in pegmatites, less frequently in quartz veins. The east part of Minas Gerais state is the richest locality. The colour of tourmaline recovered here is not always constant, as hinted at by its commercial names, such as 'Brazilian emerald', or 'Brazilian chrysolite'.

Other world-famous occurrences are in Madagascar and in the Urals. Tourmalines from Madagascar are varied in colour and frequently

316 Tourmaline — schorl — Bobrůvka (Czechoslovakia); 10 × 8 cm.

317 Tourmaline — schorl — Dolní Bory (Czechoslovakia); crystals 7 × 1.5 cm.

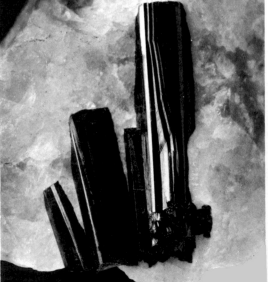

318 Tourmaline — San Pierro (Elba, Italy); crystals up to 2 cm.

resemble the Brazilian variety. A golden-yellow colouring is typical of the Madagascar stones. The Soviet deposits in the Urals, Murzinka and Sverdlovsk areas are rich in tourmalines, which occur there with amethyst, topaz and beryl in drusy cavities of coarse-grained granites. Generally they are a rich red (known as 'Siberian rubies'), but they also occur in other colours. Other tourmaline deposits are in the Republic of South Africa, Australia, Burma, Kashmir and Bengal. The largest tourmaline crystals of gem quality are found today near Muiiane in Mozambique. Their most magnificent specimen (rubellite), 42 cm long, is in the museum in Lourenço Marques. The largest fluvial deposits are in Sri Lanka. Dolní Bory in Moravia and Chudleigh and St Austell (Cornwall) have noted deposits of the black schorl variety.

As precious stones, rubellites are most in demand, especially those which resemble the ruby with their intensive and deep colour. Verdelites are also very popular, in fact all other attractively coloured transparent varieties make truly delightful precious stones. Cut stones which are vividly striped are considered a curiosity; schorls are ground occasionally into jewels for mourning. The dark, transparent varieties of tourmaline are used fairly extensively in optics.

281

**Pyroxenes** (319—324) are a group of important rock-forming minerals, in association with quartz,

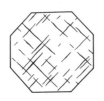

Highly complex silicates mainly of magnesium, iron, calcium and sodium.

felspar, mica or amphiboles. Pyroxenes closely resemble amphiboles. These two groups of minerals are very alike in chemical composition, shape of crystals and in colour. With a little practice it is, however, quite easy to tell them apart. Pyroxenes and amphiboles both have a fairly good cleavage, but the cleavage angle of the pyroxenes is approximately 90°, whereas that of the amphiboles is 120°. Their name is based on the Greek 'pyr' — fire, and 'xenos' — foreign, after old and completely inaccurate beliefs, that they did not originate through fire, but appeared in lava completely by chance.

Pyroxenes occur chiefly in igneous and some metamorphic rocks. Where they appear in large proportions, the rocks are classed as pyroxenic. There are, for instance, pyroxenic gneisses, pyroxenic syenite, pyroxenic granulite and pyroxenic hornstone. The truly basic igneous rock (poor in quartz), composed almost entirely of one or several pyroxene varieties, is called pyroxenite. The pyroxene varieties are named according to their various characters, for instance bronzites, diopsides, dialogites, etc.

The majority of pyroxenes have no practical significance, and only a few serve as decorative and precious stones. But they are important in theoretical practice, chiefly because they form artificially during the melting of the rocks.

Orthorhombic pyroxenes — enstatite, bronzite and hypersthene — form the so-called isomorphous series, where the magnesium silicate and iron silicate continually substitute each other. **Enstatite**, named after the Greek 'enstates' — a resistant, because of its heat resistance, occurs chiefly in massive form; the crystallized form is rare and usually imperfect. Often enstatite forms a large part of some basic rocks, either of igneous rocks or serpentines formed through their decomposition. Good crystals are found mainly in Norway. Artificially it occurs in slag. Enstatite always contains an admixture of iron; as a pure magnesium silicate it appears only in meteorites.

On account of the bronze-like colour and lustre, a variety of pyroxene is called **bronzite**, a transitory member of the enstatite-hypersthene group. It is always massive, usually forming granular or fibrous aggregates. Kraubat in Styria, Austria, is the chief locality; bronzite is found there in serpentines. It is also worked there as a precious stone, on account of its pearly metallic lustre.

Enstatite:
Magnesium silicate, $Mg_2(SiO_3)_2$, orthorhombic. H.5.5; Sp. gr. 3.15–3.176; usually greenish to dark green, grey-white, yellowish, brownish, with vitreous lustre, sometimes pearly or slightly silky on cleavage planes; S. colourless.

Bronzite:
Silicate of magnesium and iron, $(Mg,Fe)_2(SiO_3)_2$, orthorhombic. H. 5.5; Sp. gr. 3.2–3.4; commonly greenish-brown, with bronze lustre; S. colourless.

Hypersthene:
Silicate of magnesium and iron, $(Fe,Mg)_2(SiO_3)_2$, orthorhombic. H. up to 6; Sp. gr. 3.5; usually black-brown, black-green to black, with vitreous lustre, with coppery sheen on some planes; S. colourless.

Diopside:
Silicate of calcium and magnesium, $CaMg(SiO_3)_2$, monoclinic. H. 6–7 (varies in different directions); Sp. gr. 3.3; usually green to colourless, darker when iron is present; vitreous lustre on fresh planes; S. colourless.

**Hypersthene** is usually massive, in granular foliaceous form, of good cleavage; rarely does it occur in crystal form. It originates in gabbros, norites (basic igneous rocks), in erupted volcanic rocks and in cordierite gneisses. The norite variety, formed almost exclusively of hypersthene, was named hypersthenite. The gemstone varieties of hypersthene are known from the Labrador peninsula in eastern Canada. Cut and polished as bronzite, it also has a similar bronze-like brownish lustre.

Monoclinic pyroxenes, in contrast to the orthorhombic ones, do not form a continuous isomorphous group, but intermingle and replace each other in many different ways. It is therefore understandable that there are a great number of varied monoclinic pyroxene types, which can form through the intermingling of a wide variety of mixtures.

**Diopside** (319) (from the Greek 'dis' — double, and 'opsis' — opinion, because at one time mineralogists had two varied opinions about its crystals), forms short-columnar crystals or granular aggregates. It is a typical mineral of contact rocks. Diopside sometimes forms the dense contact rock malacolite, or diopside erlan, named after the deposit Erla on the German side of the Erzgebirge. The Indian diopside variety,

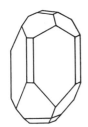

319 Pyroxene — diopside, cut — India; largest 14.0 mm in diameter; weight 13.98 carat.

320 Pyroxene — ornamental jadeite — Tibet; 6.5 × 4.2 cm.

321 Pyroxene — kunzite — Pala (California, USA); larger crystal 6 cm long.

which exhibits asterism (star-like rays being transmitted) is particularly popular as a gemstone.

The rather tenacious **jadeite** (320) was used a great deal in the past for making tools. It is usually massive, and microscopically fibrous. There are occurrences where metamorphic rocks are primarily formed by jadeite, and they are called jadeitites. The colourful type of jadeite from China and Tibet has long been used as an ornamental stone. Many objects of art and rings are made from it.

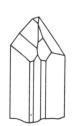

Lithium-bearing pyroxene — **spodumene** — is obtained from North American coarse-grained granites; in Dakota crystals measuring up to 12 metres have been recovered. **Kunzite** (321, 322), a transparent variety of a beautiful pink colour, is of gem quality. The colouring is caused by the admixture of manganese. This precious stone is particularly sought after and valued in America, where it occurs in the vicinity of Pala in California.

The most common and widely spread monoclinic pyroxene is **augite** (323, 324), which occurs in short-columnar crystals, either embedded or loose in some erupted igneous rocks. In this form it is found in the Central Bohemian Highlands, in Italy and in Norway. The crystals of augite are commonly perfectly bounded and sometimes twinned. They are also often flattened into a tabular shape. At times they can be found in the mould in the vicinity of their original parent rock. Like the other black-coloured minerals, augite used to be considered a variety of the black tourmaline, schorl, as late as the 19th century. It was given its present name by the German mineralogist and geologist A. G. Werner (1750–1817); it is derived from the Greek 'auge' — lustre. This mineral's highly complicated chemical composition was determined only much later. Augite is the most widespread of all the members from the pyroxene group. It is relatively hard and brittle. It is almost imune to acids.

Augite alters easily to limonite and clayey minerals. With heat it alters to chlorite, epidote and biotite, and with contact metamorphosis to the finely fibrous amphibole — uralite.

**Omphacite** (from the Greek 'omfax' — unripe, thus called because it was originally taken to be granular actinolite) resembles augite in chemical composition. It occurs massive, and is grass-green. When with garnet, it forms a rock called eclogite. The principal producers is Saualpe in Carinthia (Austria) .

Another mineral much like augite in appearance and chemical composition is **egirine** (after the idol Aigir), or **acmite** (from the Greek 'akme' — the point, on account of its pointed crystals); when the two minerals occur together, the mixture is called **egirinaugite**. Microscopic amounts develop commonly in phonolites and other rocks. It is found in macroscopic form in basic igneous rocks of Långesundfjord, Norway, on the Kola Peninsula (USSR), in Kangerluarssuk, Greenland, in Magnet Cove, Arkansas and in Serra Tingua, Brazil.

**Diallage** resembles bronzite in form, colour and lustre. It is an important rock-forming mineral of gabbros and serpentines.

322 Pyroxene — kunzite, cut — Brazil; 21.1 × 14.5 mm; weight 28.29 carat.

323 Pyroxene — augite — Arendal (Norway); crystals 2 cm.

324 Pyroxene — augite — Lukov (Czechoslovakia); large crystal 2 × 2 cm.

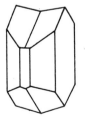

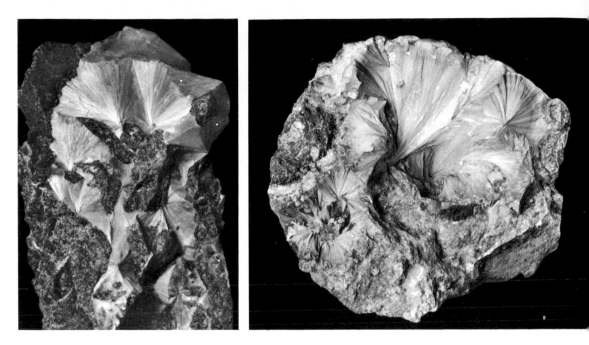

325 Pectolite — Želechov valley (Czechoslovakia); actual size of detail 6 × 5 cm.

326 Pectolite — West Paterson (New Jersey, USA); 9 × 7 cm.

**Pectolite** (325—326) was described in 1828 by the German mineralogist F. Kobell, from basalt deposits of northern Italy. He named the stone after the Greek 'pektos' — complex, and 'lithos' — stone. Pectolite occurs in the form of finely fibrous radiating aggregates, composed of individual needles, on rare occasions ending in crystal faces, at the same time forming compact fillings of cavities and cracks in rocks. It is very similar to other white fibrous silicates, such as natrolite, tremolite or wollastonite (calcium silicate, $CaSiO_3$), for which it used to be mistaken. As a mineral it is not particularly abundant.

Complex silicate of calcium and sodium, $Ca_2NaH(SiO_3)_3$, triclinic. H. 5; Sp. gr. approximately 2.8; white, grey with silky lustre; S. colourless.

The main deposits are: Monte Baldo on the east shore of Lake Garda (Italy), Bergen Hill near West Paterson in the state of New Jersey (USA), Želechov valley near Libštát in the foothills of the Krkonoše Mountains (Bohemia), southern Scotland and the surroundings of Långhorn in Sweden. Pectolite is not practically significant, but its beauty makes it very popular with collectors, especially pectolite from Paterson (USA), Scotland and Bohemia, where magnificent specimens can be found. The Swedish occurrences are most unusual and resemble asbestos.

**Wollastonite,** named in honour of the English chemist, W. H. Wollaston (1766—1828) resembles pectolite in appearance and chemical composition. It is a triclinic calcium silicate $Ca_3H(SiO_3)_3$, which forms white finely fibrous to foliated acicular aggregates. In contrast to pectolite it occurs relatively abundantly as a typical mineral of contact metamorphic limestones. The chief producers: Ciclova, Romania; Pennsylvania, USA; Quebec, Canada; Pichucalco, Mexico. In places wollastonite constitutes the chief mineral of the rock masses. Such wollastonite rocks develop through contact metamorphism of carbonaceous rocks with the admixture of quartz.

287

**Rhodonite** (327) was established as a new mineral in 1819 and named from the Greek 'rodon' — a rose, on account of its rosy colouring, which darkens and grows grey to black with oxidation. This happens because a crust of magnanese oxides (chiefly pyrolusite, psilomelane, wad and their admixtures) are deposited on rhodonite's surface during oxidation. Commonly it is massive; the crystal form is rare. Rhodonite is chiefly found in deposits of manganese iron ores. In chemical composition it resembles pyroxenes.

Silicate of manganese and calcium, $CaMn_4(SiO_3)_5$, triclinic. H. 5.5–6.5; Sp. gr. 3.4–3.68; commonly pink, dark pink to red, turns black with weathering; vitreous lustre, pearly on cleavage planes; S. colourless.

Beautiful large pink pieces occur mainly in the Urals (Maloye Sedelnikovo) and in Cummington, Massachusetts, USA, where solid blocks of quality material, weighing up to 50 kg each, are recovered. The Cummington rhodonites are most popular of all, because of their blood-red colouring. Other noted deposits are in Långban and Pajsberg in Sweden (pajsbergite), Franklin (New Jersey), Butte (Montana) and California, USA, and Australia. The compact granular massive rhodonite is a highly suitable raw material for the manufacture of many kinds of ornamental objects, especially in the USSR, the USA and Australia. From all these deposits the mineral which is veined with black manganese oxides is the most popular of all the varieties. But the chief importance of rhodonite is as manganese ore, for it contains up to 42 per cent of this metal.

327 Rhodonite — Franklin Furnace (USA); 12 × 7.5 cm.

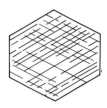

328 Microphoto of amphibole enclosed in grandiorite — Kozárovice (Czechoslovakia).

**Amphiboles** (328—333), like other black minerals, were originally mistaken for tourmaline (schorl). The German miners called amphibole 'Hornblende', because

*Highly complex basic alumino-silicates of magnesium, iron, aluminium, calcium and other elements.*

it is hard to break and as tough as a horn. At the beginning of the 19th century this mineral was put with other minerals into the amphibole group. But even this title brings one's attention to its resemblance to other minerals, for 'amfibolos' in Greek means 'ambiguous'.

Amphiboles are rock-forming minerals, which closely resemble pyroxenes (e. g. augite). They are distinguishable by the different angle of their cleavage planes (amphibole approximately 120°, pyroxene 90°). Amphiboles most commonly form irregular clusters of granular, columnar or fibrous composition, embedded in rocks. In other instances it is found in the form of columnar, smallish crystals. These minerals are widely spread in nature, and occur in many igneous rocks and crystalline schists. Among the members of the amphibole group are the industrially import-ant asbestos, and also some precious stones.

*Actinolite:*
*Complex basic silicate of cal-cium and magnesium with iron, $Ca_2(Mg,Fe)_5(OH)_2(Si_4O_{11})_2$, monoclinic. H. 5–6; Sp. gr. 2.9 to 3.2; green with vitreous lustre; S. white.*

**Nephrite** (330), a form of **jade**, is the toughest known mineral, widely used as an ornamental stone, especially in China. It is either a variety of **actinolite** (329) or **anthophyllite** amphiboles. Its exceptional toughness is caused by penetration of minute fibres of these two members of the amphibole group. Some magnificent specimens come from New Zealand

*Anthophyllite:*
*Complex basic magnesium-iron silicate, $(Mg,Fe)_7(OH)_2(Si_4O_{11})_2$, ortho-rhombic. H. 5.5–6; Sp. gr. 2.9 to 3.2; usually brownish-green to green, with pearly to vitre-ous lustre on some planes; S. white.*

and central Asia. They were soon made use of in China, where the stone is very popular and highly prized. The first occurrences of nephrite were in the range of Kuen-Lun, where it was mined in numerous, but today, com-pletely exhausted and forgotten mines. Only the boulders in alluvials still yield nephrite of good quality. Scandinavia is worth mentioning as an European deposit; from this region, nephrite boulders have been carried away by a glacier to northern Germany.

Hornblende:
Amphibole poor in silicon dioxide, containing high quantities of trivalent iron and tetravalent titanium, monoclinic — pseudohexagonal. H. 5–6; Sp. gr. 2.9–3.4; intensively black, with vitreous lustre; S. grey-green to brown-green.

**Basalt amphibole (hornblende)** is an interesting variety, found sometimes in fairly large crystals of either short- or long-columnar form, usually terminated by three faces, similar to a rhombohedron. They are usually perfectly bounded. Many rocks which are poor in silicon dioxide yield this mineral, especially igneous rocks of the Tertiary Period, in which it forms predominantly in tuffs and basalts. The beautiful crystals found in the Bohemian deposits on the Vlčák Hill near Stříbro, and in the Central Bohemian Highlands, chiefly in the vicinity of Lovosice, are world-famous.

**Common amphibole** is chemically parallel to basalt amphibole, but contains less iron. Both these minerals are markedly distinguished from the other minerals of the amphibole group by containing aluminium. The most reliable method of telling the difference between the two amphiboles is to examine the colour of a thin section; basalt amphibole is usually brown, whereas common amphibole bears various shades of green. In rocks both are remarkably like augite. Common amphibole is widespread in nature as a rock-forming mineral. It forms coloured components of many igneous rocks, and occasionally of metamorphic rocks. Amphibolites, which are schistose rocks almost entirely composed of common amphibole, can be very widespread among crystalline schists.

329 Amphibole — actinolite — Sobotín (Czechoslovakia); 9 × 7 cm.

330 Amphibole — nephrite — New Zealand; 10 × 6 cm.

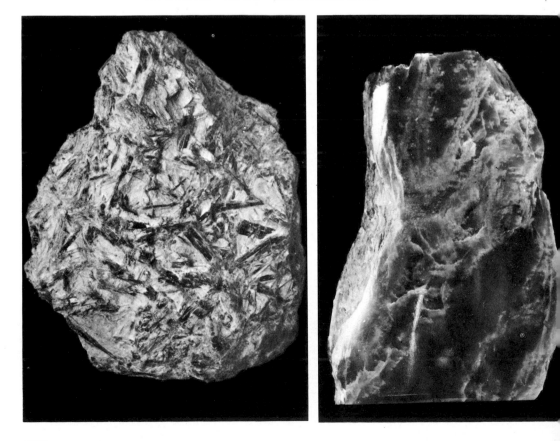

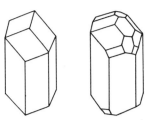

331 Amphibole — Vlčák (Czechoslovakia); 6 × 4 cm.

332 Amphibole — Lukov (Czechoslovakia); crystals 2 cm.

333 Amphibole asbestos — Graubünden (Switzerland); 11 × 8 cm.

291

**Prehnite** (334) was first brought to Europe by a Dutch colonel Prehn in 1783 from the Cape of Good Hope, South Africa. On account of the radiating, varied shades of colour, the stone was taken to be either an emerald variety, or olivine, and was called accordingly either 'capemerald', or 'capchrysolite'. Later when the same mineral was discovered in European deposits, it was thought to be a zeolite. Because of its appearance, it was referred to as 'yellow radiating zeolite'. Not until 1789 was it established to be a distinct mineral by the German mineralogist and geologist, Abraham Gottlob Werner, who was a professor at the Mining Academy of Freiberg in Saxony.

Complex basic silicate of calcium and aluminium, $Ca_2Al_2(OH)_2Si_3O_{10}$, orthorhombic. H. 6–6.5; Sp. gr. 2.8 to 3.0; usually greenish-white, with vitreous lustre. S. white.

Prehnite crystallizes in tabular form, often rounded in botryoidal or radiating masses. It occurs as a secondary mineral of aluminium-calcium silicates, chiefly felspars. Found most commonly in cavities of rocks, such as amphibolites and other crystalline schists, it is often accompanied by other secondary drusual minerals, like analcite and natrolite. The main deposits are found in: Le Bourg d'Oisans, Dauphiné in the French Alps, Scotland, Italy, Paterson in the USA, and Peru. Occasionally used as a gemstone, prehnite is usually cut into the cabochon shape.

**Bavenite** (named after the Baveno deposit in Italy) resembles prehnite, but is far more uncommon. It is an orthorhombic complex basic silicate of beryllium-aluminium-calcium, which occurs in radiating crystal sheafs composed of fine white needles. Generally it originates through the decomposition of beryl.

334 Prehnite — Scotland; actual size of detail 7 × 6 cm.

335 Apophyllite — Ústí nad Labem (Czechoslovakia); crystals 2 cm.

**Apophyllite** (335) was defined as a mineral by the French mineralogist and crystallographer René Just Haüy. He named it from the Greek 'apo' — detached, and 'fullon' — leaf, because it flakes with heating, which is due to water evaporation. This characteristic is also typical of zeolite, for which apophyllite used to be mistaken.

Complex hydrous silicate of calcium and potassium, with fluorine, $KCa_4F(Si_4O_{10})_2 . 8H_2O$, tetragonal. H. 4.5–5; Sp. gr. 2.3 to 2.4; white, greenish, pink, with vitreous lustre, strikingly pearly on cleavage planes; S. white.

Apophyllite occurs in needle-shaped, columnar or tabular crystals, or in the form of lamellar aggregates of perfect cleavage. The cloudy white opaque variety is called albite (from the Latin 'albus' — white). Apophyllite occurs relatively abundantly in druses in cavities of basaltic rocks, phonolites (basic intrusive igneous rocks) and similar formations. The most important deposits are on the Mariánská Hill in Bohemia, in the surroundings of Kaiserstuhl in West Germany, in Iceland, in Bergen County, New Jersey, and in Brazil, which is known for the occurrences of particularly large crystals. As yet, apophyllite has no practical significance.

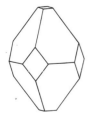

**Zeophyllite** is a similar hydrous silicate (from the Greek 'zein' — bubble over, and 'fullon' leaf); it is a triclinic hydrous basic calcium silicate with fluorine $Ca_4F_2(OH)_2Si_3O_8.2H_2O$. This mineral is also called 'the micaceous zeolite', has a perfect cleavage and forms colourless, seemingly hexagonal plates. It occurs rarely in cavities of igneous rocks in Březno and Radejšín by Litoměřice, Bohemia.

293

336 Pyrophyllite — Bedin (Carolina, USA); aggregate 13 cm.

337 Talc — Vernéřovice (Czechoslovakia); 10 × 7 cm.

**Pyrophyllite** (336), whose name is based on the Greek 'pyr' — fire, and 'fullon' — leaf, for it tends to break into leaves in the fire, is very similar to talc. It is found mainly in China and in Carolina, USA. The compact varieties of pyrophyllite (agalmatolite) have been used from days of old for the carving of statuettes and decorative objects.

Complex basic aluminium silicate, monoclinic. H. 1-2; Sp. gr. 2.2-2.3; mostly white, grey, greenish, yellowish, with greasy lustre. S. white.

**Talc** (337) received its name from the Arabic language. The name of the compact to massive variety — **steatite**, is of Greek origin. It was a familiar mineral long ago. The so-called potstone (mixture of talc and chlorites) was chiefly used in the past for the manufacture of pots; the attractively coloured varieties of talc have always been a popular material for making ornamental objects.

Basic magnesium silicate, $Mg_3(OH)_2Si_4O_{10}$, monoclinic. H. 1; Sp. gr. 2.7-2.8; commonly white, whitish to greenish (with a small amount of iron), greasy lustre; S. white.

Talc forms greasy, perfectly cleavable, flexible scales and plates. Steatite (soapstone) frequently forms part of schistose rocks. Talc is abundant in some deposits of crystalline carbonates, such as magnesites; in rocks rich in magnesium and in crystalline schists. The USA is predominant in the world's production of talc, with deposits in New York state (carbonate rocks), in Vermont and Virginia. Europe's largest deposits are in Italy and Austria. Talc is today chiefly used as a heat-resisting raw material, for instance in the manufacture of fire-resistant ceramic material.

338 Mica — muscovite — Bobrůvka (Czechoslovakia); 8 × 6 cm.

339 Phlogopite — Middletown (Connecticut, USA); 5 × 4.2 cm.

**Micas** (338 — 340) represent a group of minerals which can be split up into individual, thin, elastic plates. This characteristic made them highly usable for many practical purposes for over three centuries. Their perfect cleavage and transparency made them a suitable material for window panes.

Highly complex basic alumino-silicates of magnesium, potassium, aluminium, iron, sodium and other elements, monoclinic.

Micas are very widely spread. They occur as basic components of many rocks, such as pegmatites (especially the clear muscovite variety). They are also abundant in other igneous rocks and crystalline schists. Micas are important rock-forming minerals. According to their colour they are divided into light- and dark-coloured micas.

Muscovite:
Basic aluminosilicate of aluminium and potassium with fluorite, $KAl_2(OH,F)_2AlSi_3O_{10}$. H. 2–2.5; Sp. gr. 2.78–2.88; usually colourless, less frequently brown, pale-green, yellow, with pearly lustre; S. colourless.

**Muscovite** (338) is the most common light mica, and occurs mostly in the form of transparent plates. In other instances this mica forms only very fine scales (sericite), or entirely massive formations. Muscovite crystals are extremely rare. In the fine, scaly form muscovite is white, with a pearly lustre; the massive variety is greenish. The largest muscovite plates are found in the Urals, and in South Dakota, New Hampshire and North Carolina (USA).

Phlogopite:
Complex basic aluminosilicate of magnesium and potassium with fluorine. H. 2.5; Sp. gr. 2.7–2.9; usually reddish brown, somewhat metallic lustre; S. colourless.

**Phlogopite** (339) represents the transition between light- and dark-coloured micas. Most frequently it occurs in granular limestones of crystalline schists and in contact limestones. Its deposits are much rarer than those of muscovite, though the production is of great importance. The main producers are Canada (Templeton) and the USSR (east Siberia).

295

Of the dark varieties, whose colour is caused by the presence of iron, **biotite** is the most common. Most frequently it forms imperfect crystals, which come chiefly from Monte Somma on Vesuvius and from Canada (Ontario, Quebec). The dark micas are not very resistant and, in contrast to muscovite, they decompose easily. Usually they change into chlorites.

**Lepidolite** (340), or **lithium-mica**, usually appears in fine to coarse scales. Most frequently it forms aggregates in pegmatites. It was discovered for the first time near Rožná, Moravia. The beautiful pink and green variety occurs there with pink tourmaline and topaz. The pink variety of lepidolite is more abundant than the green. The chief localities of its occurrence are Penig (Karl-Marx-Stadt district, East Germany), Alabashka in the Urals (USSR), Hebron (Maine), and California (USA) and Cornwall (England).

The current world production of micas is approximately 50,000 tonnes per year, the chief producers being the USA (about 50 per cent), and India (23 per cent). Light-coloured micas have an extensive practical significance, for they are fire-resistant and do not decompose in acids. They are used in electrical engineering for their insulating properties, and are also very useful in optics. Their low resistance against weathering makes the dark micas less useful for technical purposes. Ground biotite is sometimes used in special paints. Lepidolite is used for the manufacture of lithium and sometimes of the admixed caesium. Lithium is employed as an additive to various alloys, particularly with lead and aluminium. Exceptionally it is used as a precious stone; cut into the cabochon shape, full advantage is taken of lepidolite's pearly lustre.

340 Mica — lepidolite — Rožná (Czechoslovakia); 8 × 6 cm.

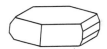

**Chlorites** (341—342), named from the Greek 'chloros' — green, are very akin to the micas, for they often form through their weathering. They resemble the micas with their perfect basal cleavage, but in contrast to them, are not elastic. If a chlorite flake is bent, it does not spring back to its original position as a flake of mica does. Their commonest occurrence is in metamorphic and metasomatic rocks. Some magnificent crystals are found in the Apennine Alps on the Swiss-Italian border and chlorite used to be called pennine; it usually occurs in small, bluish crystals, grouped in druses.

Highly complex basic alumino-silicates of magnesium, aluminium and iron, whose composition alters considerably, monoclinic — pseudohexagonal. H. approximately 2; Sp. gr. approx. 2.6–3.3 (allegedly up to 4.8); usually greenish with various shades, even black; pearly or vitreous lustre; S. colourless.

**Chamosite** (341), discovered in 1820 in southern Switzerland by a French chemist, P. Berthier, belongs to the chlorite group. Chamosite forms oolitic (grain-like) aggregates of green colour with bluish shades; in this it differs from the closely allied **thuringite** — a green-coloured chlorite named after the deposits in the area which used to be known as Thuringia (now in East Germany). Both chamosite and the less common thuringite originate in rocks by deposition from water. They are found in France and Bohemia. They are important ores of iron.

**Cronstedtite** is also included in the chlorite group. Named in honour of the Swedish mineralogist, A. Cronstedt, it is a basic complex ferroferric silicate, which crystallizes in fine black-green to black pyramids. It originates through the alteration of pyrite. This comparatively rare mineral occurs at Wheal Maudlin in Cornwall, England, in Conhongas do Campo, Brazil and in other localities. The most noted European deposits are in Romania and Bohemia.

341 Chlorite — chamosite — Baba (Czechoslovakia); 8 × 5 cm.

342 Chlorite in quartz — St Gotthard (Switzerland); actual size of detail 2 × 1.5 cm.

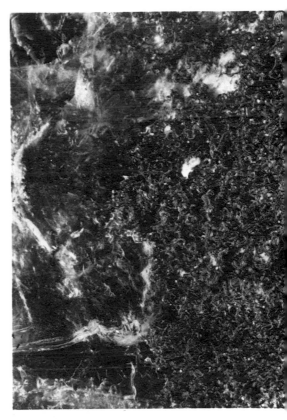

**Serpentine** (343) was already familiar in ancient Rome, where the Latin word 'serpens' — snake, gave the mineral its name, for it was the common belief that it could be used as a remedy against snake bite. Even fibrous serpentine (**asbestos**) was known to the ancient world. According to historical information, it was then mined on Cyprus. Asbestos's fire-resistant character did not then attract much attention, and was not fully appreciated even much later. Objects and materials made from this mineral were considered more a curiosity than items of any practical significance until the 19th century. Russia was the exception, for during the reign of Peter the Great (1689 — 1725), serious tests with asbestos were carried out in an effort to make the most of the rich asbestos deposits in the Urals. The Lizard in Cornwall is also rich in asbestos.

Complex basic magnesium silicate, $Mg_6(OH)_8Si_4O_{10}$, microcrystalline, monoclinic. H. 3–4; Sp. gr. 2.5–2.6; various shades of green and yellow, also reddish, striated or speckled, with vitreous lustre, very prominent after polishing; S. white.

Serpentine occurs either in the form of **chrysotile** (from the Greek 'chrysos' — golden, and 'tilos' — fibre), i.e. finely fibrous serpentine, or in tabular form, **antigorite**. Rocks formed chiefly by these two serpentine forms are called serpentinites. Balangero near Turin in Italy is the most important locality in Europe for deposits of serpentine-asbestos; the richest world deposits are in Canada (e.g. Thetford, Quebec). Because of the finely fibrous structure, chrysotile is the most important of all types of asbestos, i.e. the finely fibrous silicates, used in production of fire-resistant articles. Apart from being made into fireproof asbestos fabrics it is used for roofing (transite, for instance), asbestos cardboard, and as an insulation material against heat and electricity. The scrap material is used in various paints. Serpentine asbestos is a finer variety than amphibole asbestos. Massive serpentine is often used as building material. The precious serpentine, with a particularly beautiful colour, is used as an ornamental stone.

The nickel variety of serpentine is called **garnierite**. It is extracted as an important ore of nickel from large deposits in New Caledonia.

343 Serpentine asbestos — chrysotile — Nová Ves (Czechoslovakia); actual size of detail 8 × 5.5 cm.

344 Chrysocolla — Chuquicamata (Chile); actual size of detail 20 × 9 cm.

**Chrysocolla** (344), which in Greek means 'the golden lime', was mentioned as a known mineral by the Greek philosopher, surgeon and natural scientist Theophrastus (c. 372 — c. 287 BC), who was a pupil and follower of Aristotle. Chrysocolla was already then used as an ornamental stone. Strangely enough it was not mineralogically described until 1968.

Hydrated silicate of copper, $CuSiO_3 + nH_2O$, crypto-crystalline. H. 2—4; Sp. gr. 2—2.2; commonly blue-green, blue, brown to black, with greasy vitreous lustre; S. greenish-white.

Chrysocolla usually occurs in thin coverings or encrustations, which resemble malachite in colour. It is distinguishable by the blue shadow and can be found in the oxidation zone of copper ores. This mineral originates as a secondary product of chalcopyrite and other sulphide compounds of copper. Limonite and opal often make it impure, or it is permeated by malachite. These were the very admixtures which made the exact identification of chrysocolla's chemical composition so extremely difficult. The main deposits are in: Romania, Horní Rokytnice in the foothills of the Krkonoše Mountains in Bohemia, Eilat in Israel, in the malachite deposits near Gumeshevsk and Nizhniy Tagil in the Urals, Chile, Arizona (USA), and Cornwall (England). Chrysocolla is not a particularly important copper ore. As a precious stone it is used only exceptionally, and then chiefly in Israel and in the USSR.

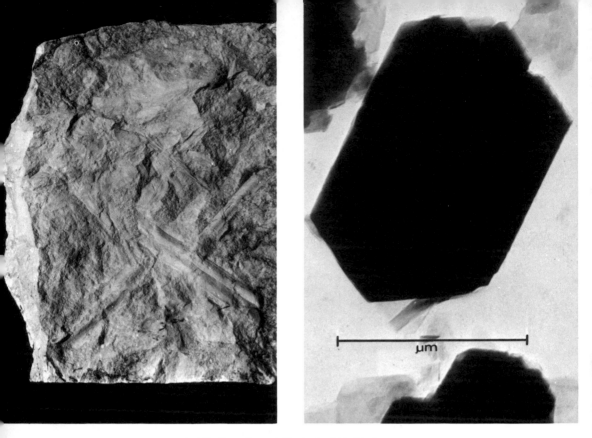

345 Kaolin — Javorná (Czechoslovakia); 11 × 8 cm.

346 Picture from electron microscope of pseudohexagonal form of kaolinite from kaolin deposit near Nepomyšl (Czechoslovakia).

**Kaolin** (345 — 346) is a name of Chinese origin, for the citizens of ancient China were working this stone by the 6th century. They ground it, mixed it with water and washed it, dried the fine deposits and pressed them into porcelain. For a long time they managed to keep the manufacturing process a secret, also the whereabouts of the kaoline deposits, which are in the present Kiang-Si province. They were successful in remaining the sole manufacturers of porcelain till the 18th century, though porcelain was brought into Iran by the 10th century and into Europe by Dutchmen in the 15th century. The Venetians and the French tried in vain to learn the secret of its manufacture. Success came at last in 1709 to the German J. Böttger.

Kaolinite:
Complex basic aluminosilicate, $Al_4(OH)_8Si_4O_{10}$, monoclinic or triclinic. H. 1–2; Sp. gr. 2.6; usually white or yellowish, with pearly lustre, greasy on massive aggregates; S. white.

Kaolin is a white clay, whose main constituent is the mineral **kaolinite**. Kaolinite is usually lamellar to earthy, and crumbly, with a greasy feel. There are other aluminosilicates present in kaolin. This mineral resulted from the decomposition of silicates rich in aluminium, especially felspars, in granites and in arkoses, which are sediments composed of quartz and felspar. The most noted deposits are St Austell in Cornwall in Great Britain, the surroundings of Karlovy Vary in Bohemia, Pennsylvania, Georgia and South Carolina (USA). Today only the very best kaolin varieties are used for the manufacture of porcelain. Those of a lower standard are used as fillers of paper stock, in paint manufacture, etc.

**Bentonite**, named after the deposit at Fort Benton (USA), resembles kaolinite in chemical composition. It is used in metallurgy and in the manufacture of bleaching clays.

**Sepiolite** (347) was well known to our forefathers under its alternative name, meerschaum. It was used for making the bowls of tobacco pipes, but these belong to the past. The sepiolite used came mostly from Turkey and Bosna. Sepiolite, in appearance and in its porous absorbent nature, resembles 'sepium' or cuttlebone, which is the internal calcareous shell of a marine cephalopod. The name of the mineral originates from this, and was suggested in 1847 by the German mineralogist and geologist E. F. Glocker, who also managed to define sepiolite's somewhat inconstant chemical composition.

Hydrous basic magnesium silicate, $Mg_4(OH)_2Si_6O_{15} \cdot 2H_2O + 4H_2O$, crypto-crystalline, orthorhombic. H. 2–2.5; Sp. gr. 2; commonly white, grey-white, yellowish, with dull lustre; S. white.

Sepiolite is commonly massive; its distinctly fibrous varieties are comparatively rare. It occurs as a secondary mineral in irregular masses in serpentines. In the form of porous tubers it is also found in sediments which have resulted from the alteration of serpentinites. When the pipe production was at its height, the main producer of sepiolite was Eskisehir, west of Ankara in central Turkey; other deposits are in Bosna, Yugoslavia, where now the extraction is negligible, in Spain, Greece and Moravia. Sepiolite is used as an insulating material, or an absorbent; it is also used in toothpaste manufacture, and as a substitute for soap, and in the making of decorative and ornamental objects.

347 Sepiolite — Nová Ves (Czechoslovakia); 9 × 6 cm.

**Sodalite** (348) was found in the ruins of the Bolivian city of Tiahuanaca, hence it must have been a fairly abundant precious stone. However, it was only defined as a distinct mineral in 1811 by the British mineralogist T. Thomson. Sodalite most commonly forms compact or granular aggregates, and on rarer occasions cleavable crystals, often twinned. Usually it is found in igneous rocks poor in quartz, such as nepheline-syenites. Quite frequently it is an important rock-forming mineral, where it is often a substitute for felspar. It is particularly abundant in sodalitites or sodalitolites, which are volcanic rocks, whose only light-coloured component is sodalite.

Complex aluminosilicate of sodium, with chlorine, $Na_8Cl_2(AlSiO_4)_6$, cubic. H. 5.5; Sp. gr. 2.3; commonly blue to rich blue, greenish to deep green, greenish-grey, grey, white, colourless, with vitreous lustre on planes, greasy on fracture; S. blue-white.

The main deposits are Ditrau in Romania, Serra de Monchique in Portugal, the volcanic formations of Monte Somma (Vesuvius), Monti Albani near Rome in Italy, Rieden in West Germany, Miass in the Urals, Ontario in Canada, Cerro Sapo in Bolivia, Greenland and the USA. In microscopic form sodalite occurs in various trachytes, phonolites and basalts. It is used today as a semiprecious stone and is usually ground to lenticular or tabular shape. Only the darker sodalites are popular, because they resemble lazurite in richness and shade of colour. The Canadian sodalites fill these requirements, and sometimes also those found in the USA and the Urals.

Sodalite often occurs with lazurite, and possibly with other related aluminosilicates, of which there are quite a number. The best known felspathoid species is **haüynite**, named in honour of the founder of scientific crystallography, a Frenchman called R. J. Haüy. In contrast to sodalite it does not contain chlorine, but sulphur and an admixture of calcium. It has a higher degree of hardness and specific gravity than sodalite.

348 Sodalite — Timmins (Canada); 9 × 6 cm.

349 Lazurite — Buchara (USSR); 9 × 6.5 cm.

350 Lazurite — Lake Baikal (USSR); 8.5 × 5.5 cm.

351 Lazurite — Canada; 9 × 8 cm.

**Lazurite** (349—351) or **lapis lazuli,** received the name from its Arabic name 'lazaward'. In Latin 'lapis' means stone. It is most probable that in ancient times lazurite was taken for a variety of sapphire. This makes it almost impossible to establish when it was first used as a precious stone. It is, however, quite indisputable, that it was always very popular as a decorative stone.

Sodium aluminosilicate with sulphur, $Na_8S(AlSiO_4)_6$, cubic. H. 5; Sp. gr. 2.38—2.9; usually mauve-blue, blue, mauve, or greenish-blue, with vitreous lustre; S. blue.

There are similar difficulties with information about the first known deposits of lazurite. Though many ancient works mention 'lapis lazuli', the localities of its occurrences and of its sales seem to be confused. But lazurite's first known deposits were without doubt in Asia and Chile.

The earliest authentic descriptions of lazurite can be found in the works of Marco Polo (c. 1254—1324). The son of a Venetian merchant, he travelled with his father and uncle across western and central Asia into China, where he stayed 17 years. He entered the service of Kubla Khan, who sent him on vital state missions, during which he travelled through many lands until then completely unknown to Europeans. This was when Marco Polo became well acquainted with lazurite deposits in the region of the upper reaches of the river Oxus (today's Amu Darya) in Afghanistan. It seems that even the ancient Egyptians transported lazurite from these deposits, to be carved as amulets, representing the sacred scarab beetle. The inhabitants of the Orient, especially of China, carved many ornamental objects from lazurite found in the local deposits. Though the deposits in question had been mined in ancient times, beautiful examples of 'lapis lazuli' are still being found there in the form of irregular masses and lenticular shapes in marble.

Lazurite occurs chiefly compact to granular, but the name lapis lazuli is mainly reserved for the compact lazurite. In this case the name is not really applied to a mineral in the correct sense of the word, but to a rock, whose components are clearly distinguishable under a microscope. This usually is a mixture of white-banded dolomite, light-coloured pyroxenes and amphiboles, minerals of the sodalite type, lazurite and grains of pyrite. The parent rock of lazurite is most commonly metasomatized stone.

The main occurrences of lazurite are, apart from Afghanistan and Chile, the southern shores of Lake Baikal and Monte Somma, Vesuvius, where it is found in erupted igneous rocks and in volcanic tuffs, similar to Monti Albani near Rome. These deposits have no real practical significance. In the Baikal area the most noted deposits are today in the riverbed of the river Slyudyanka in the eastern Sayan Mountains, where it is found in the form of large boulders. The deposits in the Andes Mountains of Chile also yield masses of gigantic sizes, which strongly resemble the lazurites from Afghanistan. But they are generally of a lighter colour and mostly mixed with a substantial amount of limestone. There are more recent occurrences in the San Antonio Canyon of southern California, but even these are not of a particularly high quality. Similar occurrences of the dark blue lazurite in the form of fine veins and grains are known also from the Italian Hill in the Sawatch Mountains, Colorado.

Lazurite was and is popular as a decorative and precious stone. It is used chiefly as a gemstone for rings, brooches and necklaces. When powdered, lazurite was at one time used as paint (the natural ultramarine).

352 Nepheline — Vesuvius (Italy); crystal 2.5 × 1.5 cm.

**Nepheline** (352) was first defined in 1800 by the French mineralogist René Just Haüy and named according to the Greek 'nefele' — cloud, because with acids it dissolves and clouds. Some ten years later the German chemist M. H. Klaproth, professor of chemistry at Berlin University and the discoverer of uranium, studied nepheline in great detail. He was chiefly concerned with the exact analysis of minerals, and paid particular attention to the gemstone variety of nepheline with its delicately white changing lustre. He confirmed that this variety, which he called **eleolite**, originated through the clouding of nepheline, through its slow process of cooling of the potassium element. Nepheline forms imperfect columnar crystals. In the form of microscopic crystals it is often a constituent of some igneous rocks, chiefly phonolites and basalts. Crystals visible to the naked eye are rarer, and known primarily from syenites. The main deposits are in Monte Somma (Vesuvius), Löbauerberg in Saxony, Norway, the Khibiny tundra on the Kola Peninsula and Miass in the Urals (USSR), also Arkansas (Magnet Cove) in the USA. It is possible to use nepheline as a source of aluminium. It is also used in the glass industry and only the eleolite variety is used as a gemstone.

Sodium-potassium alumino-silicate, $KNa_3(AlSiO_4)_4$, hexagonal. H. 5.5–6; Sp. gr. 2.60–2.65; usually colourless, white, blue-grey, green, brick-red, with vitreous to greasy lustre; S. white.

305

**Natrolite** (353) was first identified in 1824 from the best known deposit, Mariánská Hill in Bohemia. Natrolite crystallizes in long columns or slender acicular needles, frequently grouped in radiating clusters. The delicate slender needles are generally 3—4 cm long, but sometimes the columns are shorter and chubbier, ending in a low pyramid. The crystal cross-section is almost square, so natrolite could be easily mistaken for a mineral of the tetragonal group (it is pseudotetragonal). Its crystals grow on walls of cavities, where they form clusters, druses or spherical aggregates. Crystals are commonly transparent or translucent. It fuses in a candle flame and the water escaping as steam makes it swell (a characteristic of zeolites). The massive form is rare. Natrolite is abundant, and occurs as a secondary mineral in cavities of phonolites, basalts and other rocks. It is the most common mineral of the zeolite group.

Hydrous aluminosilicate of sodium, $Na_2Al_2Si_3O_{10} . 2H_2O$, orthorhombic. H. 5–5.5; Sp. gr. 2.2–2.4; white, colourless, yellowish to reddish, with vitreous lustre, silky in finely fibrous aggregates; S. colourless.

The main deposit is in Bohemia in the phonolites on Mariánská Hill; other occurrences are near Eisenach in Germany, in the Auvergne in France, Iceland, Nova Scotia, Canada. The compact natrolites from Württemberg (West Germany) and from Iceland are the only variety used as a gemstone. When cut into the cabochon shape, the Icelandic variety is known for its attractive and changeable lustre; the natrolites from Württemberg are mostly used for making small ornamental articles, often with the parent rock. Natrolite has no other practical use so far. Collectors are very keen to have it in their possession, especially specimens from the classic deposits in Bohemia and Iceland, where they occur in association with other minerals commonly found in cavities of basalts and phonolites, especially with zeolites.

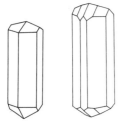

353 Natrolite — Zálezly (Czechoslovakia); actual size of detail 12 × 9 cm.

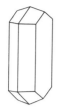

354 Scolecite — Berufjord (Iceland); 15 × 8 cm.

**Scolecite** (354) is, apart from clear calcite with its double refraction, the greatest mineralogical wonder of Iceland. It occurs there in the form of wondrously beautiful crystallized agglomerates in a number of hidden cavities and cracks in the extensive covers of young volcanic rocks which form almost the whole surface of Iceland. The most perfect examples come from the surroundings of Teigarhorn near Berufjord on the eastern shore of the island.

Hydrous calcium aluminosilicate, $Ca_2Al_2Si_3O_{10}$ . $3H_2O$, monoclinic — pseudorhombic. H. 5–5.5; Sp. gr. 2.2–2.4; colourless, white to grey, with vitreous lustre, pearly on cleavage planes; S. colourless.

The name is derived from the Greek 'skolex' — worm, because when heated using a blowpipe, it bends over. This really is a characteristic of all zeolites, hydrous silicates, which give off water when heated, which causes an alteration of their shape.

In appearance scolecite is identical to the clear or white natrolite. Like other zeolites, it occurs in cracks and drusual cavities of igneous rocks rich in cal cium and aluminium, especially in basalts and phonolites. More rarely scolecite forms in cracks of other rocks, such as granites, syenites, porphyries, porphyrites, labradoritites and crystalline schists. The main deposits are Teigarhorn near Berufjord in Iceland, Table Mountain near Golden, Colorado (USA), and especially the lava sheets of the volcanic Deccan highlands in India. Many magnificent examples of scolecite have been found there, especially in the mountains Bhor Ghat, east of Bombay, where under rail tracks and in tunnels numerous cavities have been uncovered with specimens which are now a part of all the most important museum collections. Scolecite as yet has no practical significance, but its beautiful crystals fascinate all lovers of the mineral world with their delicacy and their perfect shape.

307

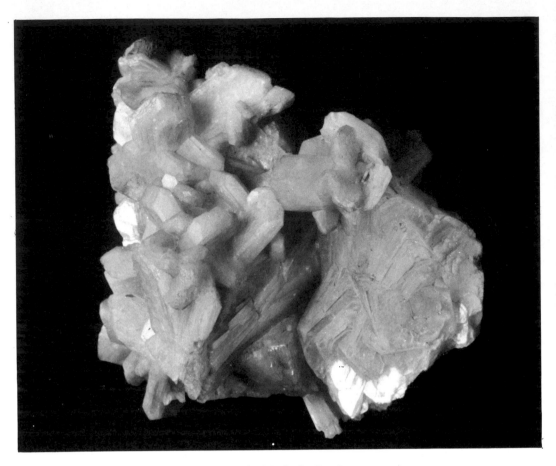

355 Heulandite — Berufjord (Iceland); 11 × 8.5 cm.

**Heulandite** (355) was established as a distinct mineral of the zeolitic group in 1822, when it was defined by the British crystallographer and mineralogist, H. J. Brook, and named after the secretary of the British Geological Society, Henry Heuland. Often misunderstanding occurs in connection with the name heulandite, for in earlier works this mineral is also called 'stilbite', which still leads to confusion, especially with the closely allied desmine, another member of the zeolitic group.

Hydrous calcium alumino-silicate, $CaAl_2Si_7O_{18}$ . $6H_2O$, monoclinic — pseudorhombic H. 3.5–4; Sp. gr. 2.2; white, brownish, reddish, with pearly lustre on cleavage planes; S. colourless.

Heulandite forms tabular crystals of perfect cleavage, also cup-shaped and scaly, radially grouped aggregates. Sometimes it is clear and colourless, or white, but more often intensively brick-red, which is due to the presence of goethite scales. Found chiefly in melaphyres, basalts, porphyrites and similar rocks, it is particularly frequent in their vesicles and cracks. The main deposits are Oberstein in West Germany; the Fassa valley in northern Italy (magnificent scarlet crystals); Kozákov in Bohemia; Strzegom in Polish Silesia; Berufjord in Iceland; the Faeroes; West Paterson near New York; Canada; the surroundings of Bombay in India. Heulandite so far has no practical significance.

Another mineral of the zeolite group — **laumontite** (named after the French mineralogist G. Laumont, who discovered it in 1785) — is a hydrous silicate of calcium and aluminium $Ca(AlSi_2O_6)_2$ . $4H_2O$. Most frequently it occurs in cracks of rocks, for example at Wolfstein in Bavaria.

308

**Stilbite** (356—357) is another member of the zeolitic group. The name is based on the Greek 'stilbein' — to glitter, and used to be applied to all zeolites with strong pearly lustre and scaly appearance. Now the name is only used by the British and the Germans, whereas in all other countries the name **desmine** is used, which is derived from the Greek 'desme' — sheaf, because of the characteristic appearance of clusters of its small crystals, which are grouped in the shape of a sheaf, and are radially lamellar to globular. Its apparently tabular crystals are always in twinned combinations. These develop by interpenetration of individual simple crystals, thus imitating orthorhombic symmetry. Stilbite is a secondary mineral crystallized from hot solutions, which flowed after an eruption through cracks to the surface, at the same time decomposing rocks in their path. This is why it is found in cracks of rocks.

Stilbite occurs in basalt and other similar rocks, in melaphyres and in amphibolites. The main deposits are: Kilpatrick in Scotland, Giant's Causeway in Ireland, the Fassa valley in Italy, Strzegom in Polish Silesia, Kozákov in Bohemia, Poona in India, Iceland, where it forms crusts on calcite beds, and in the Faeroes. Desmine has as yet no practical use, but some crystals are extremely beautiful, such as those which come from the Deccan plateau (Poona), southeast of Bombay, where they occur in extensive lava sheets in the form of large, grey-white or reddish crystals, and in sheaf-like aggregates, which are considered the most beautiful in the world.

Hydrous calcium alumino-silicate, $CaAl_2Si_7O_{18} . 7H_2O$, monoclinic. H. 3.5–4; Sp. gr. 2.1–2.2; colourless, whitish, white, greyish, yellowish, reddish, honey-coloured, less commonly brick-red; lustre strongly pearly on cleavage planes, otherwise vitreous; S. colourless.

356 Stilbite (desmine) — Berufjord (Iceland); 21 × 18 cm.

357 Stilbite (desmine) — Berufjord (Iceland); actual size of detail 7 × 6 cm.

**Mesolite** (358) was defined as an individual mineral in 1816 by the German mineralogists, Fuchs and Gehlen. Another member of the zeolite group, it was named from the Greek 'meson' — centre, and 'lithos' — stone, for, in chemical composition, it is an intermediate between natrolite and scolecite, and is, basically, a mixture of them both, roughly in the ratio of 1:2.

Complex hydrous alumino-silicate of calcium and sodium, $Na_2Ca_2(Al_6Si_9O_{10})_3 \cdot 8H_2O$, monoclinic — pseudorhombic. H. 5; Sp. gr. 2.29; white to colourless; vitreous lustre, but fibrous aggregates are fairly silky; S. white.

Mesolite occurs in the form of delicate, acicular to hair-like crystals of perfect cleavage, and in fibrous or compact aggregates. Sometimes it is earthy to powdery. A secondary mineral, it is chiefly found with natrolite in cavities of basalts and similar younger extrusive igneous rocks. Mesolite is fairly abundant, more so, perhaps, than pure natrolite. Its occurrences are basically identical with those of natrolites. Localities: Central Bohemian Highlands, Pflasterkaute near Eisenach in Germany (Thuringia), the island of Skye in Scotland, Antrim in Northern Ireland, Nova Scotia (Canada), the Faeroes. Mesolite has no practical use. The largest mesolite crystals, needles up to 15 cm long, come from Teigarhorn in Iceland. The most magnificent specimens come from localities in New Jersey, USA, where they form unusually thin white needles. But the occurrences in the state of Oregon are the most unusual of all; there the fibrous crystals form powerful, continuous masses with a visibly fluctuating lustre.

358 Mesolite — Horní Hrad (Czechoslovakia); 10 × 10 cm.

359 Thomsonite — Pihel (Czechoslovakia); 12 × 8 cm.

**Thomsonite** (359) was identified as a separate member of the zeolitic group by the British crystallographer and mineralogist H. J. Brook, who named it after his friend T. Thomson. Sometimes it is also called **comptonite**. Thomsonite usually forms columnar or tabular crystals, generally grouped in radial clusters and bunches. In other instances it occurs in the form of reniform or globular aggregates with a radially fibrous composition and drusual surface.

Complex hydrous alumino-silicate of calcium and sodium, $NaCa_2(AlSiO_4)_5 . 6H_2O$, orthorhombic. H. 5–5.5; Sp. gr. 2.3 to 2.4; almost colourless, white, greyish, yellowish, greenish, reddish, with vitreous lustre, pearly on cleavage planes; S. colourless.

Thomsonite is found fairly abundantly in cavities of phonolites and basalt rocks, more rarely in pegmatite veins. The main deposits are in: Eisenach in East Germany, Kilpatrick near Dumbarton in Scotland, Vinařická Hora near Kladno in Bohemia, Colorado and Lake Superior in the USA. The best examples of thomsonite are today found in the northern region of New Jersey State in the USA. These are white-coloured star-shaped aggregates up to 5 cm in diameter, which in shape resemble stilbite. The radial aggregates composed of fine fibres occur in Nova Scotia (Canada). Thomsonite is occasionally used as a precious stone and then ground to a lenticular shape. Thomsonite from Good Harbor Bay on Lake Michigan is mainly used for this purpose. Here thomsonite occurs as globular fillings of vesicles and cavities, and the crystals are sometimes as much as 3 cm long. They appear chiefly on the lake shores, where they are found in the form of pebbles. They are usually radially arranged and are concentrically striated, with alternating stripes of milky-white, yellow and green colour.

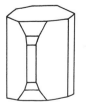

311

360 Chabazite — Iceland; 12 × 8 cm.

**Chabazite** (360—361) was identified in 1788 from deposits in Oberstein in the Rhineland. The name comes from a supposed Greek work 'habazios' — which in the early 19th century was believed to have been used for an unknown, beautiful mineral in an ancient Greek poem. However, it is probable that 'habazios' is actually a mistaken spelling of 'halazios' — hail. Crystals of this mineral, up to 2.5 cm long, were later found near Řepčice in Bohemia.

Hydrous calcium alumino-silicate, $Ca(AlSi_2O_6)_2 \cdot 6H_2O$, rhombohedral — pseudocubic. H. 4.5; Sp. gr. 2.1; white, red, with vitreous lustre; S. colour-less.

The rhombohedral chabazite crystals are often intertwinned in regular combinations, forming lenticular shapes, so-called **phacolites** (361). Apart from deposits already named, there are also occurrences near Strzegom in Polish Silesia, in Iceland, Canada, Chile and Australia.

361 Chabazite — phacolite — Richmond near Melbourne (Australia); crystals 10—15 mm.

312

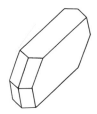

362 Phillipsite — Vinařická Mountain (Czechoslovakia); actual size of detail 5.5 × 5.5 cm.

**Phillipsite** (362) is yet another member of the zeolite family, discovered in 1825 and named after the British geologist and mineralogist J. W. Phillips (1773 – 1828). Though one of the most abundant zeolites, phillipsite occurs only in the form of minute, always intricately intertwinned crystals, or in globular, centrally radiating aggregates. Phillipsite is a monoclinic mineral, but the development of its twins imitates the cubic system.

Complex hydrous aluminosilicate of calcium and potassium, $KCaAl_3Si_5O_{16} \cdot 6H_2O$, monoclinic; H. 4.5; Sp. gr. 2.2; colourless, whitish, white to yellowish, with vitreous lustre; S. colourless.

Phillipsite is found in basalts and phonolites as a secondary mineral together with other zeolites. It also appears, but less abundantly, in some other young extrusive igneous rocks. The main deposits are Stempel near Marburg in Essen, Capo di Bove near Rome, the vicinity of Česká Lípa in Bohemia, and Richmond in New South Wales, Australia. Phillipsite has no practical use but the distinctly intertwinned crystals are very popular collection items.

Similar penetration twins are developed by yet another mineral from the zeolite group — **harmotome**, a hydrous silicate of barium and aluminium $Ba\,Al_2Si_6O_{16}.6H_2O$. The shape of the twins gave the mineral its name (from the Greek 'armos' — joint, and 'tomos' — cutting, because the twins are formed by crystals which are separated by grooves).

313

**Analcite** (363) was first described at the end of the 18th century from deposits in the basalt rocks of the Cyclopean cliffs, northeast of Catania in Sicily. It is possible to charge it electrically by rubbing. Analcite forms either individual, or drusually arranged crystals, which are usually imperfectly developed. It is also granular, compact and earthy. The most typical occurrences are drusual cavities of rocks, basalts especially.

Hydrous sodium aluminosilicate, $NaAl(SiO_3)_2 \cdot H_2O$, cubic. H. 5.5; Sp. gr. 2.2–2.3; colourless, white, slightly brown, with vitreous lustre; S. colourless.

Main deposits are: Fassa in Trento and Vesuvius (Italy), Andreasberg in the Harz Mountains, Čáslav in Bohemia, Blagodat in the Urals, Bergen Hill in New Jersey. Analcite has no practical use but the attractive crystal specimens are popular collector's items. The original deposit in Sicily yields the best examples; it is still a rich locality and in the basalt cavities, clear and often quite large crystals, which sometimes completely fill the cavity, can be found. The north Italian deposit at Val di Fassa, Trento, yields the most exquisite examples of beautiful rose-shaded crystals, which are a joy to any collector.

363 Analcite — Alpe di Seis (Italy); 12 × 10 cm.

364 Leucite — Vesuvius (Italy); crystal 2 cm.

**Leucite** (364) was already known to natural scientists in the second half of the 18th century. Their interest was caught by leucite crystals embedded in the Vesuvian lava, which in shape resembled garnet crystals. They differed, however, in their grey or white colour. This is why leucite was originally taken for a special variety of the garnet, which had lost its colour through the action of acid volcanic gases, or through the direct influence of the 'volcanic fire'. It was not till much later that it was proved that this is an entirely different mineral. A detailed chemical analysis showed that this was the first mineral known to contain a substantial amount of potassium, which till then was thought to occur exclusively in plants.

Potassium aluminosilicate, $KAl(SiO_3)_2$, cubic or tetragonal. H. 5.5–6; Sp. gr. 2.5; commonly greyish-white, with vitreous lustre; S. colourless.

The crystals of leucite are of a regular form, with twenty-four faces. These are not really crystals in the true sense of the word, but pseudo-cubic forms of tetragonal leucite, which develops from the leucite under normal temperatures. Leucite is found mainly in young, volcanic rocks. The main deposits, apart from Vesuvius, are the shores of Lake Laacher, West Germany, São Paulo in Brazil and Magnet Cove in Arkansas (USA).

315

365 Felspar — adularia — Kandy (Sri Lanka); pieces approx. 2 cm.

**Felspars** (365—376) were named from the German 'feld' — field, because during their weathering and decomposition, they free great numbers of plant nutrients, such as potassium, and in this way enrich the soil. They are therefore vital components of plant nutrition.

Aluminosilicates of potassium or sodium-calcium, monoclinic or triclinic.

There are two important classes of felspars — potash felspars and soda-lime felspars (plagioclase). The potash felspars crystallize either monoclinically (orthoclase) or triclinically (microcline). The conditions for the crystallization of orthoclase and microcline are almost identical, so it is understandable that a seemingly uniform crystal can be partly orthoclase and partly microcline. Both the minerals occur in interesting varieties and commonly contain an admixture of other elements, particularly sodium. Plagioclases form a continuous isomorphic series with soda-felspar (albite) and lime-felspar (anorthite) as end members. The two best known intermediate members, which are also the most important ones, are oligoclase, with a predominance of soda-component, and labradorite, with a predominance of lime.

Felspars are usually highly cleavable, compact, granular or granularly lamellar, or they form tabular or short-columnar crystals. Intergrowths of two individuals are common and there are many types of different twinning combinations, usually named according to their main occurrences. Orthoclase, for example, develops Carlsbad twins, Manebach twins, or Bavena twins. Such intergrowths are less common with plagioclase minerals, though they have repeated twinning according to two laws — the albite law and the pericline law. With most felspars this twinning is often repeated and gives a series of fine lamellae, which are

316

frequently microscopic. All felspars are most commonly found in the form of twins or multiple twins and intergrowths, where several individual crystals grow into each other.

Felspar crystals usually have a perfect cleavage in two directions which are perpendicular or almost perpendicular to each other. In this they differ from other minerals. They are comparatively hard, and usually light coloured.

Felspars are the most important rock-forming minerals. They are the dominant part of igneous rocks, such as granites and pegmatites, but also of far more basic rocks poor in quartz, such as diorites, melaphyres, porphyrites, basalts, tephrites, etc. In acidic rocks potash felspars are dominant; in basic rocks the soda-lime felspars (plagioclases) dominate. But felspars are also a vital part of a number of crystalline schists and sedimentary rocks, especially of arkoses (sandstone composed chiefly of quartz and felspars). Some cavities of igneous rocks, particularly pegmatites and granites, contain crystals of potassium felspars as much as one metre in length. They weather rather easily and are the basis of the origin of the rich kaolin deposits which are abundant throughout the world.

Felspars, and particularly orthoclase, are important industrially. They are used chiefly in the glass and ceramic industries, in the manufacture of porcelain, porcelain-glass and enamels. According to their various uses, it is therefore also possible to divide the felspars into ceramic felspars with a higher content of silicon dioxide (this applies mainly to potash felspars), and glaze felspars, which contain a high proportion of sodium oxide. Felspar is an important ingredient in the manufacture of glass, for it supplies not only aluminium oxide but also potassium oxide and sodium oxide.

Felspars are also utilized during the manufacture of abrasives and

366 Felspar — orthoclase — Strzegom (Poland); crystals up to 2 cm.

materials of extreme hardness used for grinding and smoothing; they are also used as cleansing agents (fillers of soap), and in dentistry. Felspar powder is used in fertilizers. Some felspar varieties are cut as precious or ornamental stones, especially adularia, amazonstone and labradorite.

The greatest extraction of felspars is in the USA (over a quarter of a million tonnes of felspathic raw material per year). The deposits are centred mainly in the states along the Atlantic coast (North and South Carolina, Maine), and South Dakota. Other large occurrences are in Canada, Sweden, France, the USSR, China and Great Britain. Pegmatites yield all the felspar varieties suitable for industrial uses.

**Orthoclase** (366) received its name from the Greek 'orthos' — straight and 'klasis' — cleavage, on account of its good cleavage. The thickly tabular and short-columnar crystals are commonly intertwinned. The Carlsbad twins occur most frequently and are so named because they were originally discovered in the region of Karlovy Vary (for which Carlsbad is the German name) in Bohemia, where they are still recovered.

Such intergrowth is the result of two individual crystals intertwinning according to crystallization laws along the plane which is parallel to the vertical crystal axis and the twin axis. (The two crystals are not positioned one behind the other as they intertwine, which is common in twins, but grow together side by side and mutually penetrate each other.)

Orthoclase:
Potassium aluminosilicate, $KAlSi_3O_8$, monoclinic. H. 6; Sp. gr. 2.54–2.56; usually yellow-white or pinkish to blood-red, with vitreous to pearly lustre; S. colourless.

367 Felspar — adularia — St Gotthard (Switzerland); $10 \times 9$ cm.

When Carlsbad twins are perfectly developed and fairly large, it is possible to determine the right twin from the left twin according to the side from which the second crystal has grown onto the base. The twinning of orthoclase according to the Carlsbad law is widespread in nature, and is not limited to the large twins found in granites. It is often visible even on microscopic grains of potash felspar in the basic ground mass of other rocks.

Finely tabular crystallized orthoclase with an admixture of sodium is called **sanidine** (from the Greek 'sanis' — plate). It occurs in younger igneous rocks, such as trachytes and phonolites. The optical characteristics of sanidine differ somewhat from those of other orthoclase varieties.

Pure orthoclase is completely colourless, but mostly it is white, yellow or flesh-pink. The variety **adularia** (365, 367) originates from hot solutions and is the purest of the orthoclase family. It was named after the deposits on Mount Adula near St Gotthard in Switzerland, where it occurs in cracks of Alpine rocks, just as in hundreds of other deposits in the Mont Blanc massif. Adularia is transparent, often permeated with chlorites, and sometimes displays a delicate play of colours. Its crystals,

368 Graphic (alphabet) granite — Shaitansk (Urals, USSR); 9 × 6 cm.

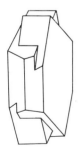

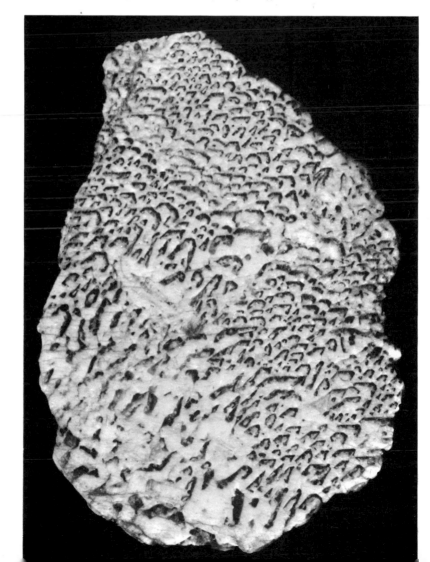

369 Felspar — microcline — Crystal Peak (Colorado, USA); crystals approx. 30 mm.

which frequently fill cracks of schists are often bounded by perfect faces. In contrast to the other felspars they have a pearly lustre.

Orthoclase is one of the most widely spread silicates in the mineral world and one of the most abundant rock-forming minerals. Not only does it represent a substantial part of many igneous rocks, but it also forms part of a great number of crystalline schists and rocks of sediment-ary origin, sandstones, etc. Here it usually forms tiny, equally distributed grains. In some granites orthoclase occurs abundantly in the form of larger crystals, superior to the other minerals of uniform grain size found in the rock. Such so-called porphyritic phenocrysts are often perfectly bounded and have the shape of flattened simple columns. Basically they are phenocrysts several centimetres long, from granite magma, which are older than the adjacent basic substance and which crystallized from as yet unsolidified melt. Their crystal bounding is therefore more perfect than the bounding of potassium felspars, which are a component of the basic fine-grained granite substance.

The most exquisite orthoclase crystals are not usually found in the basic rock, but in cavities of granites, where there is ample space to develop their shape regularly. Such coarse-grained granites, which always developed with the influence of volcanic gases, are called pegma-tites. They are the most promising grounds for mineral collectors, yielding beautiful crystals of extremely large sizes, which are commonly accompanied by potassium felspar, usually orthoclase.

During the decomposition of granite, they often fall out of the original

370 Microphoto of plagioclase felspar in quartz diorite — Krhanice (Czechoslovakia).

rock and find their way into gravels and into top soils, where they are most commonly found. The Cornish granites, Shap granites and Mourne Mountain granites are good British examples. The Carlsbad twins, whose parent rock is the porphyritic Carlsbad granite, are found in the same manner in many Bohemian deposits. The twins are, as a rule, 5 to 8 cm long; on rare occasions twins with a rough surface, measuring as much as 10 cm, have been found.

The most magnificent orthoclase crystals form in drusual cavities of granites, where they have plenty of space which they need to form regular crystallized shapes. Crystals from these cavities reach considerable lengths, sometimes as much as a metre. In some coarse-grained granites, which originated during the action of magma fluid, especially in pegmatites, where orthoclase crystallized in comparatively large sizes, it is often permeated with quartz. This is the origin of the so-called **graphic (alphabet) granite** (368), for quartz mimics the shapes of letters of the Hebrew alphabet. The title was suggested by the French mineralogist René Just Haüy, who described this penetration in great detail in 1822. Orthoclase often mixes with microcline in graphic granites, sometimes even in a single crystal.

After R. J. Haüy many other research workers and scientists explored and studied systematically the intertwinning of felspars and quartz; they were also trying to explain their true origin. Two opinions prevailed. According to the first, the intergrowth occurred through concurrent crystallization of both minerals when the temperature of solidification of the magma was reached. The second opinion disagrees with the

321

synchronous crystallization and states that the quartz constituents developed in felspars through their partial replacement (metasomatism). According to this theory, hot, circulating liquids dissolved the felspar mass and replaced it with the less soluble quartz, which was in the solutions. The dispute about these two theories still continues today. If we study in detail the intertwinning of felspar with quartz, we will see that quartz forms imperfect columns, which are often hollow, and which appear on the cleavable faces of felspar as dark-grey cross-sections; they rather resemble Hebrew letters, or the runes (the angular characters of the alphabet used by ancient Germanic peoples) and are therefore sometimes also called runite.

Alphabetical intertwinning of felspar and quartz, or the graphic structure of granites and pegmatites, is very abundant in some localities. In many pegmatites it occurs in very large pieces and at times is clearly visible in their veins. In other instances it is not visible to the naked eye and can be determined only with the aid of a microscope. It is then called a micrographic structure. This develops commonly in felspars in the basic porphyries and similar rocks.

Alphabetical striations are also fairly common in the triclinic microcline (for istance from the Ilmen Mountains in the Urals). But they are less prominent in the triclinic sodium-calcium member of the felspar family (plagioclases). If quartz penetrates into them, which happens only with acid plagioclase, such as albite or oligoclase, the striations along the cleavage planes usually resemble worms. Such a structure is called a worm-structure, or a myrmekitic structure.

Orthoclase weathers fairly easily, often to kaolinite (as in the beds of Karlovy Vary in Bohemia), or into clay with muscovite.

The best orthoclase crystals come from San Piero on the Italian island of Elba, from Murzinka in the Urals, and especially from Madagascar, where in one locality a gem variety of orthoclase occurs, beautifully transparent and yellow in colour. Velké Meziříčí in Moravia also yields well formed, nicely bounded crystals, and so do Strzegom and Jelenia Góra in Polish Silesia, also northern Italy (Baveno) and Scandinavia.

In all these localities orthoclase crystals occur in drusual cavities of granites and pegmatites. In Frederiksvärn, Norway, there are occurrences of orthoclase crystals with an exquisite play of colours. The best examples of the variety adularia come from Sri Lanka, from where they are now exported to the whole world. It is an excellent raw material and is extracted mainly in the central parts of the island and in the rich alluvial deposits on the southern shore.

The yellow, perfectly transparent orthoclase crystals from Madagascar are particularly suitable as gem-material, and they are frequently ground into cuts with many facets. The milky clouded variety of adularia, called **moonstone**, is a prized gemstone; with cutting and polishing it shines with a unique, beautiful bluish lustre of different shades. At the present time adularia from Sri Lanka is particularly popular and imported into European grinding plants.

Attractive moonstones of gem quality are today also found in many areas of the USA. Most unusual specimens come, for instance, from a new deposit on the western slopes of the Black Range near Noonday Canyon in Grant County, New Mexico. Crystals here are not of the adularia variety, but are sanidines and often reach exceptional dimensions of 30 to 60 cm. The outer part of these crystals is mostly milky white, whereas the centre is transparent. The volcanic region of New Mexico provides at the present time many other places where there are occurrences of exquisite felspar crystals with the typical moonstone play of colours. Pointed intergrowths up to 7 cm long are known from the northern region of Embudo Canyon, from the western slope of Sandia Mountain in Bernalillo County, Oregon, and from the South Canyon

district in Dona Ana County. Felspar crystals enclosed in siliceous monzonite which occur in the Burro Mountains in Grant County and in the
Luna district and in Otero, are also as large. Fine examples of Carlsbad
twins come from the monzonite porphyry in the Red River district in Taos
County. Similar, but smaller examples of Carlsbad twins are found in
substantial quantities near Ray in Pinal County, Arizona; they occur in
a pass which has been named accordingly 'the Crystal Pass'. Lovely
white adularias, called **valencianites**, which form either simple crystals
similar to the Swiss adularia, or twins reaching only exceptionally 2 cm
in length, come from the mines of the Silver City district in Owhyee
County, Idaho. Though the local stones are not ground, they are some
of the most beautiful stones to grace any collection. Moonstones also
occur among sanidine crystals in deposits in Colorado, such as in the
liparites of the Ragged Mountain in Gunnison County. These also are
often of exceptional dimensions, but are not commonly as perfectly
developed as crystals from the previously mentioned deposits.

The delicate play of colours of the cut adularia somewhat resembles
the precious opal, though the colours are not as vivid. Adularia-moon-
stones are ground into the lenticular cut, and so is the Norwegian
orthoclase — **perthite**, which is also known for its delicate play of colours,
which is similar to labradorite. This iridescence is caused by the interference of light rays with the delicate soda-felspar components present,
which are enclosed in orthoclase (so-called natronorthoclase). Orthoclases

371 Felspar  —  albite  —  Dolní Bory (Czechoslovakia); 7 × 6 cm.

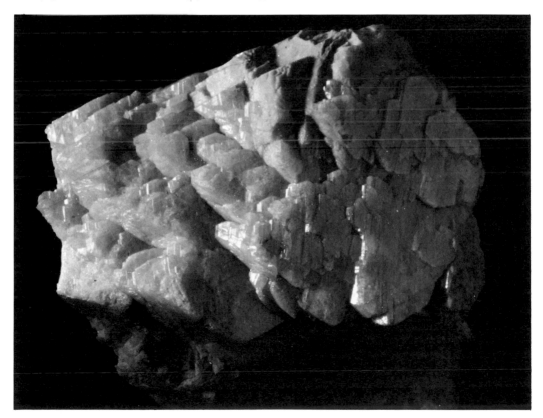

372 Felspar — pericline — Sonnblick (Austria); actual size of detail 6 × 6 cm.

which display iridescence are particularly suitable to be ground for ornamental purposes, especially to be used as stones for rings.

Orthoclase crystals from some deposits are also very popular with collectors of minerals. The Carlsbad twins are particularly desirable in this sense, especially the specimens found in Loket (which are sometimes called 'Loket twins'), and in other original localities in the surroundings of Karlovy Vary in Bohemia. Perhaps their popularity is due to the fact that they can be seen in every major mineral collection in the world. Yet these are not particularly valuable specimens, though twins which are of a substantial size and with perfectly bounded faces are not exactly numerous.

**Microcline** (369) was named from the Greek 'mikros' — imperceptible, and 'klinein' — to incline, because the angle between its cleavage planes is slightly different from a right angle. Microcline closely resembles orthoclase, with which it is often intergrown. The blue-green variety is called **amazonite**, which is really a misnomer, for the green stone found in the deposits by the River Amazon in South America is nephrite and not microcline.

Crystals of microcline are commonly large and perfectly bounded, flat columnar and very similar in appearance to orthoclase. This is because the conditions of crystallization are almost identical for both these potash felspars. Their crystals are frequently intertwinned, always according to the laws of twinning, never by chance. When greatly magnified under an electron microscope, the two minerals show no difference in their intertwinned forms.

Microcline:
Potassium aluminosilicate, $KAlSi_3O_8$, triclinic. H. 6; Sp. gr. 2.57; white to malachite green (amazonite); lustre similar to orthoclase; S. colourless.

324

Microcline occurs mostly in granites and granite pegmatites. But there are also intrusive igneous rocks, composed almost entirely of microcline, and classified therefore as microclinites. Microcline is fairly abundant in nature. The deposits most important from the mineralogical point of view are Jelenia Góra in upper Polish Silesia, Iveland in Norway, where it develops in druses of pegmatites, in Finland, Hagendorf (West Germany), Miass in the Urals and Magnet Cove in Arkansas (USA).

The occurrences of the blue-green amazonite are in Madagascar, in Amelia, Virginia and near Pike's Peak, Colorado in the USA, from where the most beautiful and most prized gem varieties are recovered. Amazonite is often found here in pegmatites and red granite. The deposits are scattered over an extensive area on the west side of Pike's Peak. The best specimens come from a large pegmatite mass near Crystal Peak north of Lake George in Teller County, Colorado. Crystals from here are green to dark blue-green (these are the most valuable); sometimes they are enclosed by white albite, which often forms star-shaped agglomerates of small foliated crystals (called clevelandite). Druses composed of felspar crystals and of the black-brown, strongly lustrous crystals of smoky quartz are particularly popular. Light brown microcline crystals in cavities of pegmatites are a common occurrence. The largest local individual crystals are as much as 40 cm long. Their centres are, however, always formed with quartz. Crystals without the quartz core reach a maximum length of 12 cm. Beautiful amazonite specimens of gem quality come from several other American deposits in Virginia. In Rutherford mine no. 1 and Morefield mine near Amelia, crystals which

373 Felspar — oligoclase — Tvedestrand (Norway); actual size of detail 9 × 6 cm.

374 Felspar — oligoclase, cut — Tvedestrand (Norway); 42 × 32 mm; weight 15.7 g.

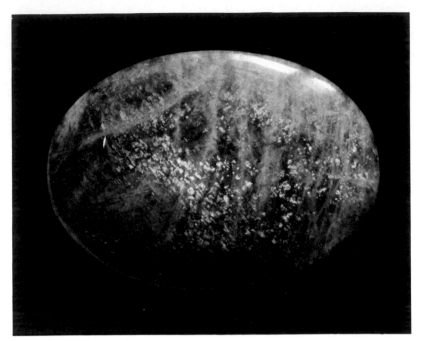

were a component of the rock and which were 45 cm long have been found. Other attractive examples of amazonite come from the noted Canadian deposit in Renfrew County, Ontario. These are, however, commonly massive, similarly to the Brazilian amazonites. There are known occurrences of perfectly bounded microcline crystals from Tawara on the island Honshu, Japan. They are up to 9 cm long, but are white and non-transparent. The richest deposit is Miass in the Ilmen Mountains in the Urals, where it occurs in coarse masses, but also in magnificent crystals.

The coarse-grained alphabet pegmatites in the Ilmen Mountains, composed of the blue-green amazonite and the smoky-grey quartz started to crystallize at a temperature of about 800°C. The gradual cooling of the melt then ensured the right conditions for the growth of gigantic felspar crystals. The regular minute design of the alphabet granite formed here at a temperature of 575°C, when the dark quartz separated from the solution. Then, at lower temperatures more irregular alphabet patterns developed and all the time larger and larger crystals of quartz were formed. One of the founders of geochemistry, a noted Soviet mineralogist A. J. Fersman, studied the mineralogy and geochemistry of felspar occurrences in the pegmatite region of the Ilmen Mountains in great detail.

According to earlier information, the average extraction per year in that locality reached two tonnes. Amazonite was mined with other local gemstones and with mica by Russian Cossacks in the 18th century, during the constant skirmishes with the local pugnacious Bashkirs and Kazakhs. There are some beautiful table-tops made from amazonite from this period, which today still grace the chambers of the Hermitage Museum in Leningrad. One of the amazonite quarries in Miass is no longer mined, but is now a nature reserve. It has been mined carefully into one giant amazonite crystal, which is alphabetically striated with quartz.

**Albite** (371) is the most abundant triclinic plagioclase felspar, and is named from the Latin 'albus' — white. It is never a completely pure felspar of aluminium and sodium, as is apparent from its chemical formula, but always contains an admixture of calcium. Albite crystallizes in smallish, tabular to short-columnar crystals, often twinned. Elongated crystals, which are twinned in a specific manner characteristic of some plagioclase (the so-called pericline law), are called **periclines** (372).

Albite occurs in pegmatites, some granites, syenites and trachytes, always as a rock-forming mineral. In the crystal form it is found chiefly in fissures of various rocks. The main localities of its occurrence are Schmirn in the Austrian Tirol, near Strzegom in Polish Silesia, Bobrůvka near Velké Meziříčí in Moravia, Tintagel (Cornwall) and the Mountains of Mourne (Ireland). The beautiful albite variety displaying iridescence (**peristerite**), found in Canada and Virginia (Bedford County), is used as a precious and ornamental stone. It is ground to the lenticular cut.

**Oligoclase** (373—374) (from the Greek 'oligos' — little, and 'klasis' — fracture, because of its low cleavage) is a coarsely granular to compact felspar. It is exceptional to find it in columnar or short-columnar crystals. Oligoclase occurs in granites, syenites, diorites, porphyries, trachytes and andesites, but also in gneisses and serpentines. It is comparatively abundant. The main deposits are Tvedestrand, Norway, and the Urals in the USSR. As a semiprecious stone it is polished and ground to the lenticular cut, especially the grey-brown oligoclase with distinct iridescence which comes from Norway.

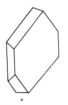

Two hundred years ago, a missionary called Father Adolf, discovered on the eastern shore of the Labrador peninsula of Canada some rather strange stones. They were in large quantity, but of a rather dull grey colour, but when they were turned the colour changed into pleasing dark blue and green hues. This discovery aroused exceptional attention among natural scientists, and after some time they identified this unusual mineral as one of the triclinic felspars of sodium and calcium (plagioclase), and named it after the place of discovery — **labradorite** (375—376).

Shortly afterwards it was discovered in large cleavable masses in other deposits, which also exhibited a beautiful change of colours, such as Kosoi Brod near Kiev in the Ukraine. Labradorite boulders were already being worked in 1781 in the famous grinding workshops of the town of Petrodvorets near Petrograd (today's Leningrad), to where they were transported from the Ukraine. The attractively coloured pieces with their marvellous change of colours were turned mainly into small orna-ments and occassionally into jewelry. The local labradorites with their yellow, orange or golden shadows were particularly prized. In no other deposits do they exhibit such perfect merging of colour with movement. The rare clear-blue stones with a greenish shadow were sometimes called 'ox eye'.

The dark play of colour of labradorite, which is characteristic of this variety, is caused by the refraction of light from the microscopic penetra-tions of small crystals of various dark minerals, which are oriented in the direction of its perfect cleavage. These 'intruders' are most commonly fine scales or needles of ilmenite, magnetite, or haematite, as has been established by a thorough examination under the microscope. As far as the lighter shadows, particularly the blue shadows, are concerned, it appears these are caused by the reflection and interference of rays of light in the fine twinned intergrowths of labradorites, which can also be determined only with the help of a microscope.

Labradorite is most commonly massive, granular, and only rarely forms tabular crystals, which are invariably very small. Such minute particles were found embedded in andesites near Roșia Montană in Romania, also in the volcanic ash and tuffs on the slopes of Mount Etna in Sicily.

375 Felspar — labradorite — St Paul (Labrador, Canada); $10 \times 4.5$ cm.

Labradorite also occurs as a basic constituent of various other igneous rocks poor in quartz, such as porphyrites, gabbros, diorites, basalts and diabase rocks, etc. Often it represents a vital rock-forming component. An intrusive rock composed almost entirely of labradorite, is called labradoritite. Labradorite also occurs in crystalline schists. It is a fairly abundant mineral.

Apart from the deposits already mentioned, among which the Labrador coast is still the largest and the most important one, Ojamo in Finland is a heavily mined and well known area. But the gem quality occurrences of labradorite are only found in the original deposit. In eastern Labrador large crystal boulders up to 60 cm in size are extracted today from anorthosites, which are rocks composed almost entirely of labradorite. Beautiful granular labradorite is also widespread throughout many localities of the New York State, particularly in Essex County.

Labradorite is without doubt the most extensively used precious and ornamental stone with a play of colours, which can be further enhanced with proper cutting and polishing. Like the very first discoverers of labradorite, we too can observe the unusual play of various shades even on stones which have not been treated. There are, indeed, only a very few minerals which display a similar play of colours at a glance (sometimes called labradorization). Today labradorite is mainly ground to the lenticular and tabular shape, or is used for the manufacture of very popular and fairly expensive decorative articles. Labradorite rock is an apposite material for use as building panels or in the construction of monuments.

The calcium felspar **anorthite**, the end member of the isomorphous group of triclinic plagioclase felspars, always contains an admixture of sodium. Most commonly it forms druses of columnar or tabular crystals, frequently twinned. Anorthite is rarer than the other plagioclases mentioned, occurring mostly as an essential component of basic igneous rocks, such as anorthosites, which is a term used for intrusive igneous rocks composed almost entirely of plagioclase. The chief occurrences of anorthite are on Vesuvius, in the Urals and in the Antilles.

Anorthite:
Calcium aluminosilicate, $CaAl_2Si_2O_8$, triclinic. H. 6; Sp. gr. 2.76; usually greyish, reddish, white to colourless, with vitreous lustre, pearly on some cleavage planes; S. white.

376 Felspar — labradorite — detail.

## Chapter 10

ORGANOLITES (ORGANIC COMPOUNDS)

The organolite group consists of natural salts of organic acids, natural hydrocarbons (i.e. compounds of carbon and hydrogen in various proportions), and fossil resins. Previously, all types of coal and petroleum were included in the organolite group, but now they tend to be classified as rocks. All the same, even now there exist many substances which stand on the boundary of the mineral system, and are counted as minerals by some mineralogists. Ozocerite for instance, belongs to this group. Others consider as minerals only the salts of organic acids, which in the form of minerals occur only rarely in the earth's crust. These are compounds dependent on the biosphere, but it is interesting to note that whewellite, for instance, has been found even in ore-veins. Organolites develop in the earth's crust in various ways. Very often minerals which, at other times, develop inorganically (sulphur, apatite, calcite, aragonite, pyrite, etc.), are produced organically.

**Ozocerite** (377) was first noticed in the sandstones near Slanic, northwest of Plojest in Romania in 1833, where it filled fissures. As it is extremely similar to wax, ozocerite was named 'mineral wax' or 'native paraffin' or 'mineral tallow'. It is a rather tough, sticky and unattractive material, which commonly forms tabular, vein-like, globular, botryoidal or reniform agglomerates. Ozocerite originates through the oxidation of petroleum deposits, when it either forms layers, or the less important fillings of rock cracks and

Mixture of hydrocarbons, amorphous. H. 1–2; Sp. gr. 1.1–1.2; light yellow, green-brown, brown to blackish, resinous lustre; S. white or lighter than surface colouring.

377 Ozocerite — Truskowiec (Poland); 9 × 8 cm.

cavities. There are extensive deposits on the coast of the Dead Sea, around lake La Brea on the island of Trinidad, on the shore of Lake Bermudez in Venezuela and in the USSR. Most of the world production of ozocerite goes into the manufacture of insulators for electric cables.

**Whewellite (378)**, though rare and unusual in its crystal form and crystalline aggregates, in origin and in chemical composition, is not a new mineral. Ancient miners saw it many times in cracks of rocks near coal seams, but they took it to be clear quartz (rock crystal), calcite or barite. In 1840 it was eventually described by the British researchers, W. H. Miller and J. Brook, probably from deposits in Transylvania, and named in honour of the British natural scientist, William Whewell.

Hydrous oxalate of calcium $Ca(C_2O_4) . H_2O$, monoclinic. H. 2.5; Sp. gr. 2.23; colourless to clear, more uncommonly whitish to greyish, with pearly lustre; S. colourless.

Later it was again discovered near Freital-Burgk and Freiberg in Saxony, and in some coal mines of the Erzgebirge (Ore Mountains). All the same it was a great surprise to the experts when whewellite was found in 1906 in the form of magnificent crystals in the cracks of the Theodor coal mine, northeast of Kladno in Bohemia. Perfectly bounded and beautifully clear crystals up to four centimetres long were found there in clusters measuring up to a decimetre. As time progressed, more occurrences and deposits were found in the Kladno area. It was ascertained that whewellite is a faithful companion of coal seams, and that it occurs fairly abundantly in their bed. Crystals from the Kladno area are some of the most perfectly crystallized examples of this particular mineral in existence.

Whewellite is a typical mineral of coal basins, but it is never found directly in the coal. It develops only in its vicinity, either directly in rocks or in globular concretions of claysiderites. Apparently it originated from solutions of calcium oxalate during coal formation, which is a simple process of leaching from bodies of plants. Whewellite crystals thus formed everywhere where the conditions were suitable. It is an interesting fact that the biggest crystals always occur in fissures of rocks, whereas in pelosiderites there are commonly small crystals accompanied by crystallized ankerite, barite, kaolinite and by tiny crystals of some sulphides. Whewellite is the most uncommon of all these minerals and is also the last one to originate. Occurrences in ore-veins are less common. Here whewellite is formed through the effects of the oxalate released from plant material on neighbouring minerals containing calcium.

Whewellite crystallizes most frequently in the shape of multifaced columns, or plates, and pointed crystals of varied appearance. Heart-shaped or butterfly-shaped twins are often seen. The crystals are usually grouped in lamellar or star-shaped aggregates. Crystalline encrustations are very common. It is distinguishable from the very similar barite by the perfect cleavage in several directions.

The main deposits of whewellite are at Kladno and at Capnic in Romania, where in 1926 a crystal almost 7 cm long was discovered; however, not a single Romanian crystal is as perfectly developed as the crystals from Kladno. Rich deposits have recently been discovered in the oil fields near Maikop in the Caucasus (USSR). Whewellite is undoubtedly the most beautiful mineral of the organolite group.

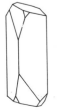

Another mineral of the oxalate group is **humboldtine** (named in honour of the noted German natural scientist, traveller, geographer and geologist Alexander von Humboldt, 1769—1859), or **oxalite**. It is a monoclinic hydrous oxalate of iron $Fe(C_2O_4) . 2H_2O$. It forms elongated, seemingly orthorhombic crystals, or is pisolitic, fibrous or earthy. The crystal form occurs at Capo d'Arco on the Italian island of Elba. The similar orthorhombic **oxamite** is a hydrous oxalate with ammonia $(NH_4)_2(C_2O_4) . H_2O$. Commonly it is yellowish-white and forms a component of guano found off the coast of Peru.

331

378 Whewellite — Kladno (Czechoslovakia); 4 × 2.8 cm.

**Mellite** is a hydrous mellitate of aluminium $Al_2(C_{12}O_{12}) \cdot 18H_2O$. The name is derived from the Latin 'mel' — honey, because of its honey-yellow colour. It occurs as white or yellow pyramidal crystals or is massive, granular or embedded in rock. It is found in sandstone and in coal. The chief producers: Artern in Halle, East Germany, and the Tula district, USSR.

**Dopplerite** is a mixture of humus acids and calcareous humates. It forms reddish-brown greasy covers on bog and peat beds, where it develops through the oxidation of humus substances, which penetrate into gaps of the organic substratum, or penetrate into cracks among wood remains, where with the evaporation of water they concentrate and harden. This mineral was first described as a 'gelatinous substance' by H. Doppler in 1849. Later it was named in his honour by the Austrian mineralogist and geologist Wilhelm von Haidinger (1795–1871).

**Fichtelite** (379) was first discovered in 1837 in the peat-bogs at Redwitz, in Bavaria, southwest of Cheb. The name comes from the German mountain range Fichtelgebirge,

Hydrocarbon, $C_{19}H_{34}$, monoclinic. H. 1; Sp. gr. 1.03; colourless to yellowish; greasy lustre; S. colourless.

where it occurred. Later, far more abundant deposits were discovered in the south Bohemian bogs, during the recovery of peat near Soběslav. There it occurs fairly commonly on timber remnants, especially on pine logs and branches, which at some time in the past were devoured by the bog. Fichtelite forms lamellar aggregates, crystalline crusts and coverings, or occasionally elongated tabular crystals. Usually they are minute, and only rarely reach 7 mm. Often they are stretched in one direction. It is a delicate mineral, of perfect cleavage, greasy to handle and easily spreadable. Other fichtelite deposits are for instance in Kolbermoor, southern Bavaria; near Sulzendeich by Oldenburg; in shaley coal near Uznach, southeast of Zürich in Switzerland; in the peats near Heltegaard, Denmark, and near Handforth in Cheshire, England. These deposits of fichtelite have as yet no practical significance. But they are proof that rare crystallized minerals can originate even during the decay and alteration of old vegetation and animal matter.

**Kratochvilite**, named in honour of the Czech petrologist J. Kratochvíl, is native fluorene $C_{13}H_{10}$. It occurs as tiny purplish scales with a pearly lustre on pitheaps in Kladno, Bohemia.

**Hatchettin**, named after the British chemist C. Hatchett, is a higher hydrocarbon of the methane series. It crystallizes as small, yellowish or white plates without definite bounding, and occurs in some coal seams, such as in the coal field of Rosice-Oslavany in Moravia.

**Asphalt** is a mixture of higher hydrocarbons. It is a collective name for solid bituminous minerals.

379 Fichtelite — Mažice (Czechoslovakia); 8.5 × 5.5 cm.

**Amber** (380—382) was one of the most prized precious stones of the primeval and early ancient times. Prehistoric man liked the stone because it was easily workable at his own particular stage of development. The soft amber became very popular, for it was easily ground, was attractively coloured and, when polished, shone with a high lustre.

Fossil resin, amorphous. H.2 to 2.5; Sp.gr. 1.0–1.1; yellow to yellow-red, with greasy to waxy lustre; S. colourless.

Necklaces made of amber beads have been found dating from the beginning of the 2nd millennium BC, in the Old Bronze Age. Simple amber ornaments were, however, made as long ago as the Old Stone Age (Palaeolithic Period), as is proved by excavations in several caves, chiefly in central and northern Europe. Amber was also used as a means of payment, possibly even in the Stone Age, and quite definitely in the Bronze Age.

Ancient Assyrians and Babylonians were very fond of amber jewelry and even the Greek poet, Homer, in the *Odyssey* praises the beautiful amber. By then amber was correctly classed as a fossil resin, as mentioned by Aristotle (384—322BC). The ancient Romans even knew that when rubbed with a piece of cloth, it picked up small particles. This was discovered in the 6th century BC by the Greek philosopher Thales of Miletus. Later, when it was proved that this was the action of electricity, the energy was given the name after the Greek expression for amber — 'elektron'.

Roman women loved to dress themselves in amber jewelry, and amber ornaments had a high value. The great distances between amber deposits and the borders of the Roman Empire made this stone expensive and desirable. It is known that Nero (AD 37—68) sent special expeditions to recover this mineral from the Baltic coast. The so-called 'amber paths', which led from the Baltic across Europe to the mighty Roman Empire are well known from that era. The main path led from the Baltic across Moravia, through eastern Austria into Italy.

Amber was an important item on the exchange market and was also greatly favoured by the scholars of Rome. Pliny the Elder (AD 23—79) discusses amber's origin and agrees with the opinions of Aristotle. Tacitus (c. 55—120) mentions how it was used by the Germanic tribes.

The popularity of amber has survived through the ages. Amber jewelry was made in the Middle Ages, and even now is still in great demand. The fact that every European language has its own ancient name for amber shows how well known this mineral must have been even then. The Greek name, 'elektron', was used for the first time in the 3rd century BC, by Theophrastus, a pupil of Aristotle. The name suggests 'lustrous metal'. Perhaps the clear lustrous yellow colour of amber, so enhanced with cutting and polishing, led Theophrastus to choose such a name.

The Latin name of 'succinum' is old too, and was the source of the later scientific title of amber — succinite. The German name 'Bernstein' is derived from the word 'brennen' — to burn, and reminds one of the easy flammability of amber. The English term 'amber' derives from Medieval Latin 'ambra' and the Arabic 'anbar', and the Slavonic term is 'jantar'.

Amber is a fossil resin of the Tertiary coniferous trees. Thick forests grew in the Tertiary Period, and resin trickled from damaged trees in the same way as it does today in our woods, but the trees of the Tertiary Period were richer in resin. It flowed from the wounds and cracks and formed small drops or larger pebbles. These peeled off the tree and dropped to the ground. In time they were covered with other sedimentations. In the meantime many millions of years passed by, until erosion by water exposed them once more.

Amber occurs as irregular nodular fragments. It is usually cloudy, but sometimes translucent, even transparent, and is delicate and exceptionally light in weight. When heated, amber's transparency improves. The largest known boulder, preserved from the Baltic Sea deposits, is in the collection

380 Amber — valchovite — Valchov (Czechoslovakia); 10 × 8 cm.

of the Natural Science Museum of the Humboldt University in Berlin. This specimen weighs almost 8 kg. It is said that lumps weighing as much as 10 kg have been found on the Baltic coast.

In some amber nodules remains of old vegetation and insects are preserved (382). Such remains from the Tertiary Period, as much as 50 million years old, became trapped and glued to the flowing resin. These finds are scientifically most important, for sometimes the remains have been preserved in the minutest detail. They have to be studied enclosed in amber, for it is impossible to extract them.

The best known amber deposit is of course on the coast of the Baltic Sea, where it appears particularly after sea storms, when it is thrown onto the shore by strong waves. The 'blue soil' of the Tertiary sandstone is amber's parent rock in that instance. The largest deposits are northwest of Kaliningrad, especially in the vicinity of Jantarnyj (former Palmnicken). Amber is also found on the shores of Poland and Germany, in Denmark, Sweden, Holland and eastern England. Baltic amber can even be found in many central European localities, south of their original deposits, chiefly in Silesia. Here it occurs in rocks which were brought during the Ice Age by glaciers from the Baltic region. These are secondary deposits.

Smaller quantities of amber have also been found in other localities,

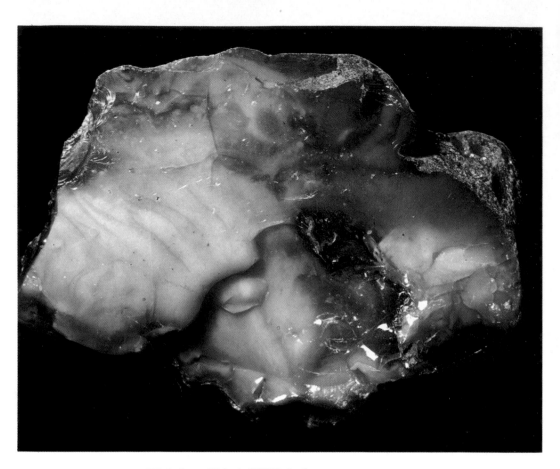

381 Amber — Klaipeda (USSR); 7 × 5 cm.

mainly in the Ukraine round Kiev and Lvov, and in Romania, where there have been occurrences of amber nodules weighing as much as 3 kg. Strikingly fluorescent blue and green nodules come from Sicily, especially from the alluvials of the river Simeto near Catania. This amber variety is therefore called **simetite**. With its dark-brown colouring it is conspicuously different from the Baltic variety. The conditions of origin in all these deposits were similar to those on the Baltic coast.

Fossil resins, which greatly resemble amber, also occur in Moravia and in the neighbouring region of Bohemia. Such resins, which are classified under separate names, differ from amber chiefly in their geological age. They originated in the Cretaceous Period.

Amber found in the Valchov region in Moravia, was named **valchovite** (380) accordingly. Waxy yellow in colour, it forms fair-sized nodules in coal and sandstone of the Cretaceous Period. In lignite coal of the same age, yet another mineral similar to amber was found — **muckite**. In contrast to valchovite it is dark brown in colour. Even a pale-yellow resin — **neudorphite** — was found later in the same coal.

Today amber is still widely used for the manufacture of various ornamental objects, though the time of its highest popularity (16th to 18th centuries) has now passed.

The amber which is recovered today is divided into groups according to quality, and then is worked in modern factories. A special lathe is used for grinding the amber and it is polished with sepiolite and with diatomaceous earth. Many commercial objects, such as cigar and cigarette holders and other smoking accessories, and various souvenirs and ornaments are made from amber. The transparent variety with the pleasant colour hues is used almost exclusively for the manufacture of facetted jewelry, such as necklaces, bracelets and rings. During the manufacture of any works of art, full advantage is taken of the local colour shades of the particular amber, such as striations, merging of colours, or cloudiness, and also of variety of form. Even the scrap from the amber raw material is not wasted. When heated under high pressure, 'compressed amber' (ambroid) is recovered.

The Sicilian simetite is valued more than the Baltic amber. The fluorescent stone, which it is possible to enhance still further by the right working methods, is particularly highly prized. On occasions, the resins from the Bohemian-Moravian border, so similar to amber, are also used as pendants or other ornamental articles.

Though the main use of amber is for making ornaments, this mineral also serves in the production of succinic acid, and succinic oils, also in the production of insulating materials for delicate electrical apparatus. Amber oils are also used in medicine.

382 Amber with insects — Balt; actual size of insects 10 mm.

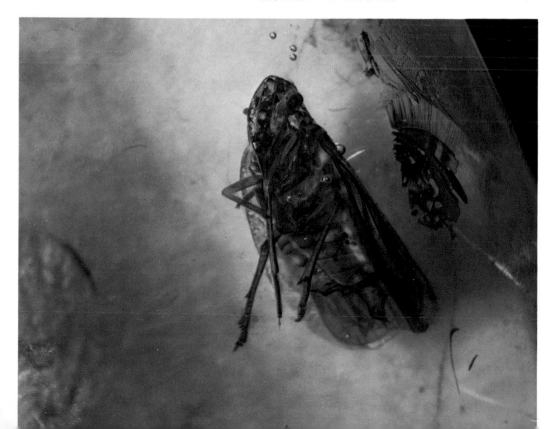

# Chapter 11 METEORITES AND TEKTITES

This book has now described all the principal members of the ten groups of minerals, i.e. inorganic natural substances which are chemically and physically homogeneous, whether they have been formed by inorganic or organic processes. Mineralogy, however, is also concerned with other inorganic natural matters, though they do not strictly fulfil the definition of a mineral. Meteorites and tektites are such substances, and are in many characteristics closer to rocks than to minerals. Their composition and origin are also exceptional, and though they do not belong to any mineral system, they are extremely interesting objects. As the study of meteorites and tektites can answer many questions regarding the composition of extraterrestrial bodies, conditions in the universe and in the upper strata of the atmosphere, and last but not least the composition of the core of the earth itself, it is natural that today scientists' interest in the subject has risen greatly.

**Meteorites** (383—386) are the remnants of an extraterrestrial body which has fallen onto our planet. Today they arouse even greater interest because of their rarity and because our knowledge of the universe has been greatly extended through space exploration. Such falls have, of course, always been the subject of mystery and even horror. In 1135, the Prague chronicle of the Canon of Vyšehrad states: 'A giant stone as big as a house fell out of the clouds on a plain in Thuringia. People living in the vicinity could hear the noise it was making three days beforehand. As it hit the earth, half of the stone became embedded in the ground and for three days it remained red hot, like steel is when taken out of the fire.' This description of a meteorite fall is somewhat inaccurate. All the same, it is most

383 Stone meteorite — Luby (Czechoslovakia); 18 × 14 cm.

important, for scientists of the Middle Ages categorically denied that stones could 'fall from the skies'.

In 1754 the Prague astronomer, Josef Stepling, observed and described 'the shower of stones' which fell in the Tábor district of Bohemia. He was the very first to state with absolute certainty that these stones fell to the earth from outer space. All the same, as late as 1790 the French Academy of Sciences in Paris voiced its displeasure and regret at hearing similar ideas, and accounts from eye-witnesses of actual meteorite falls were dismissed as imaginative stories. In 1794, the famous German physicist E. F. Chladni swept aside all the old objections with definite scientific proof.

According to some current mineralogical experiments and opinions of some astronomers, meteorites are considered to be the remnants of a large planet, which in composition and character greatly resembled the earth, and not just the core of the earth, but also the surface layers. The planet moved through our solar system between Mars and Jupiter, and then for some unknown reason it disintegrated. Its particles are apparently still journeying through space. These fragments, when they reach about 80 to 150 km above the earth, come into contact with our atmosphere, and because they are flying at the tremendous speed of 30 to 60 km per second, the air friction makes them red hot and sets them on fire. They

384 Iron meteorite — Tazewell (Tennessee, USA); 9 × 6 cm.

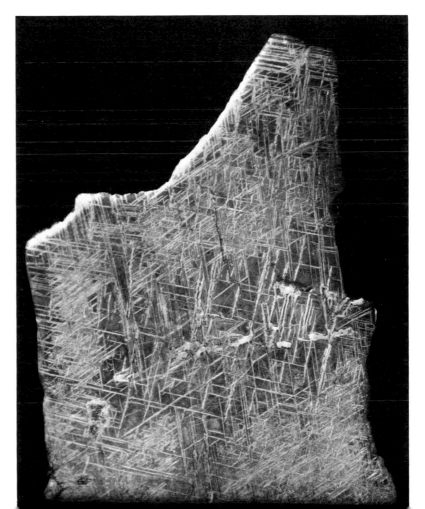

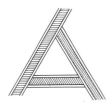

385 Pallasite — meteorite — Springwater (Saskatchewan, Canada); 26 × 14 cm.

flare up in the sky suddenly, like golden shining stars, which flash across the night sky. Only a few ever reach the earth's surface, however.

Meteorites are divided according to their composition into **iron meteorites** (384, 386) and **stone meteorites** (383). The intermediate types, formed by components of iron and stone, are very rare. Iron meteorites, which are exceptionally heavy, were the first to be found and are the most common. They are composed chiefly of iron with an admixture of nickel. As a rule, every iron meteorite contains two alloys, with varied nickel content. When the meteorite is cut, and the sawn surface is etched with acid, the lamellae of these alloys are revealed. According to their structure they are divided into octahedrites, which have the lamellae rich in nickel arranged as an octahedron, hexahedrites, with cube-like structure and ataxites, which have no structure at all.

Stone meteorites are usually smaller and very like the earth's igneous rocks. But their surface is formed by a black crust, which is created by melting under intense heat during its flight. **Pallasites** (385) are intermediate between iron meteorites and stone meteorites. Basically they are

386 Iron meteorite — Loket (Czechoslovakia); 16 × 14 cm.

composed of iron with an admixture of nickel, and olivine crystals are often found, usually rounded in shape.

The largest known iron meteorite is a boulder, which lies near a farm in the vicinity of Grootfontein in Namibia. It weighs about 60 tonnes and is guarded as a valuable natural monument. The biggest stone meteorite, which weighs over one tonne, fell to earth on 18 February 1948 in Norton County, Nebraska in the USA, as one of the components of a so-called 'stone shower', which occurs when stone meteorites break into smaller pieces during their fall, and only exceptionally hit the ground as a single stone. An example of such a collective fall of meteorites is the remarkable shower of approximately 300 meteorite stones, which struck the earth at Stonařov near Jihlava in Moravia in the year 1808.

Some beautiful pallasites have been found in several localities in North America. In 1885, for instance, a pallasite weighing over 9 kg was found in Kansas in Brenham Township. In 1931 three lumps of a similar meteorite with a total weight of about 68 kg were found in the vicinity of Springwater in Canada. Here the embedded olivine grains measured as much as 4 cm.

**Tektites** (387 — 390) are rather mysterious objects and are today the subject of much scientific interest. The best known and also the longest known are the **vltavines** (387, 388, 390) or **moldavites**, which are found in Bohemia and Moravia. They were discovered in the second half of the 18th century, and were at first thought to be olivine, then volcanic glass, or vitreous slag from old glassworks. When several other deposits of similar glass were discovered at the end of the 19th century, as far away as southeast Asia and Australia, it was generally thought that all these substances must be glass meteorites. They were given the collective name of tektites.

Astronomers, however, disagreed with the opinion that these substances were of meteorite origin, and went on to prove that glass meteorites cannot exist. Several new theories were then put forward. According to one, tektites were formed after a large meteorite had hit the earth, melting the rocks where it fell, and scattering at the same time its glass particles in all directions. Another theory states that such glass is of lunar origin, where it was formed in a similar manner when a meteorite hit the moon; the glass fragments recoiled and fell onto the earth's surface, for the gravitation of the earth is more powerful.

The origin of tektites is still shrouded in considerable mystery. The fact that they do not contain radio-isotopes of any elements show they could not have undertaken a very long journey through the universe. The recent tests on moon rocks have proved that the lunar theory of their origin is incorrect. It seems therefore more plausible that tektites owe

Silicate glass with a high content of $SiO_2$, amorphous. H. 6–7; Sp. gr. 2.3–2.4; olive-green to black-green, vitreous lustre; S. colourless.

387 Tektites — vltavines from various deposits in Czechoslovakia; largest 46 mm.

388 Tektites — vltavines — Lhenice (Czechoslovakia); larger 5 × 4 cm.

389 Tektites — billitonites — Belitung (Indonesia); 24 and 22 mm.

their origin to the earth itself. Coesite, which is fundamentally silicon dioxide and forms only under high presssure, has been found in tektites. In contrast, bubbles found in tektites contain gas with a low pressure. These conditions point to only one possible mode of origin: that, at the time of impact of a giant meteorite on the earth, the atmosphere is subjected to high temperatures and pressures, causing vacuum bubbles to form on the earth and vacuum tunnels in the atmosphere. Tektites therefore really are the altered surface layers of rocks, which through the impact were partially melted down to glass, partially evaporated and partially recondensed. Then thermal currents dispersed these particles far and wide. This is why every tektite deposit occurs around particular meteorite craters. The crater Ries near Stuttgart is the home of the Bohemian and Moravian vltavines.

Vltavines and other tektites occur as cobbles, with a noticeable coarsely wrinkled surface. They are dark-green to green-black, but show a strong green colour when subjected to direct light. They are found chiefly in loose sediments in the vicinity of České Budějovice and Moravian Třebíč. Vltavines were worn as pendants, or ground as precious stones. Even today they are a popular gemstone. Usually they are left in their natural state and are set in gold or silver. Tektites from other deposits are also used as gemstones, but their surface sculpture is not as striking as that of vltavines, so they are not as popular.

At the present time industrial mining of vltavines is being considered at a particular south Bohemian deposit, but they lack the properties of precious stones. This applies to other tektites too. They are not hard enough, damage easily and often lack the necessary lustre. They owe their popularity to their unsual shape and the mystery of their origin.

343

390 Tektite — vltavine (transmitted by light) — Netolice (Czechoslovakia); 5 × 5 cm.

A vltavine found in Switzerland has, however, recently been accepted among the gem-quality elite. During a visit, the Swiss Government presented Queen Elizabeth II with a piece of jewelry in which a vltavine is inset into platinum together with a diamond and black pearls.

Many other tektites apart from vltavines are named after their places of origin: australites, javanites, indochinites, philippinites and thailandites, even ivorites from the Ivory Coast of West Africa, and bediasites and georgianites from North America, and others.

All tektites are similar in chemical composition and appearance, but differ in shape and shade of colour. The primary difference is, however, in their geological age. Their age can be determined by radioactive investigation (the potassium-argon method) or by the splitting of the uranium core. The first test sees a change in state of the potassium isotope (kalium) to the argon isotope. The second test measures the traces left from the spontaneous decomposition of the atoms of uranium. It has been established that the North American tektites are the oldest (more than 34 million years old), and ivorites the youngest (1.3 million years). Vltavines have not been found which are more than 15 million years old.

# BIBLIOGRAPHY

*Bancroft, P.:* Minerals and Crystals. The Viking Press, New York, 1973

*Bauer, J.:* A Field Guide in Colour to Minerals, Rocks and Precious Stones. Octopus Books, London, 1975

*Borner, R.:* Minerals, Rocks and Gemstones. Oliver and Boyd, Edinburgh, 1966

*Bottley, E. P.:* Rocks and Minerals. Octopus Books, London, 1972

*Chudoba, K. F., Gübelin, E. J.:* Edelsteinkundliches Handbuch. Wilhelm Stollfuss, Bonn, 1974

*Del Caldo, A., Moro, C., Gramaccioli, C. M., Boscardin, M.:* Guida ai Minerali. Fratelli Fabbri Editori, Milano, 1973

*Desautels, P. E.:* Mineral Kingdom. The Hamlyn Publishing Group Ltd., London, New York, Sydney, Toronto, 1971

*Evans, E. K.:* Rocks and Rock Collecting. Golden Press, Western Publishing, Wisconsin, 1970

*Heavilin, J.:* Rocks and Gems. Macmillan, New York, 1964

*Irving, R.:* Rocks and Minerals. Knopf, New York, 1956

*Michele, V. de:* Minerali. Istituto Geografico De Agostini, S.p.A., Novara, 1971

*Michele, V. de:* Guida mineralogica d'Italia 1, 2. Istituto Geografico De Agostini, S.p.A., Novara, 1974

*Rutland, E. H.:* Gemstones. The Hamlyn Publishing Group Ltd., London, New York, Sydney, Toronto, 1974

*Schubnel, H.-J.:* Pierres précieuses, gemmes et pierres dures. Grange Batelière S.A., Paris, 1969

*Schubnel, H.-J.:* Pierres précieuses dans le monde. Horizons, Paris, 1972

*Seim, R.:* Minerale. Neumann Verlag, Leipzig, 1970

*Simpson, B.:* Rocks and Minerals. Pergamon, New York, 1966

*Sinkankas, J.:* Mineralogy for Amateurs. Van Nostrand Reinhold Company, New York, Cincinnati, Toronto, London, Melbourne, 1964

*Stalder, H. A., Haverkamp, F.:* Mineralien. Mondo Verlag, Lausanne, 1973

*Strunz, H.:* Mineralogische Tabellen. Akademische Verlagsgesellschaft, Leipzig, 1970

*Talent, J.:* Minerals, Rocks and Gems. Tri-Ocean, San Francisco, 1970

*Vollstädt, H.:* Einheimische Minerale. Verlag Theodor Steinkopff, Dresden, 1971

*Vollstädt, H., Baumgärtel, R.:* Einheimische Edelsteine. Verlag Theodor Steinkopff, Dresden, 1975

# INDEX OF MINERALS AND ROCKS

Roman figures refer to text pages, figures in italics indicate numbers of illustrations.

Rubellite **278**, 279, 280, 281, *314*
Rubicelle s. Spinel
Ruby 12, 13, 16, 17, 20, **104**, 106, *104, 107*
Ruby, Asteriated 105
Ruby spinel s. Spinel
Rutile 26, 31, 129, **174 — 176**, *188 — 191*

Sagenite 129, 174, 175, *188, 191*
Salt, Common s. Rock salt
Sand **38**, 122
Sandstone 9, 24, **38**, 39, 122, *29*
Sanidine **319**
Sapphire 12, 13, 16, 17, 20, **104**, *12, 105, 106*
Sapphire, Asteriated 106
Sard **139**
Sardonyx **151**, 157
Satin ore s. Malachite
Satin spar **206**
Scheelite **213**, *241*
Schorl **278**, 279, 281, *316 — 317*
Schwazite **60**
Scolecite **307**, *354*
Sekaninaite **277**, *313*
Selenite **206**, 207, *232 — 233*
Semi-opal **159**, 170
Sepiolite **301**, *347*
Sericite 295
Serpentine 99, 160, 198, 236, 240, 245, **298**, *343*
Serpentinite **38**, 298
Siderite 39, 187, **199**, *224*
Silver 12, 21, 35, **42 — 44**, 50, 61, 66, *35 — 37*
Silver glance s. Argentite
Silverstone s. Calcite
Sillimanite **250**
Simetite **336**, 337
Slate 9, **38**, *27*
Slate, Chlorite *32*
Slate, Crystalline **39**
Slate, Pyrite **37**
Smithsonite 187, **200**, 261, *225*
Soapstone s. Steatite
Sodalite **302**, *348*
Sodalitite 302
Sodalitolite 302
Sodium **190**, *207*
Spessartite **241**, *274, 277*
Sphalerite 36, 61, **62**, *56 — 57*
Sphene **259**
Spinel 13, 16, 17, 30, 31, 94, **96 — 97**, *95 — 96*
Spodumene **284**
Sprudelstein **192**, *212*
Staffelite **225**, *256*
Staurolite **257 — 258**, *292 — 293*
Steatite **294**
Stilbite **309**, *356 — 357*
Stilpnosiderite 185
Stolzite **213**

Stone meteorite **340**, *383*
Succinite s. Amber
Sulphur 34, **51 — 52**, *45*
Sylvine **90**

Tabasheer **162**
Talc 27, 28, **294**, *337*
Tantalite **180**, *194*
Tanzanite **267**, 268
Tektite **342 — 344**, *387 — 390*
Tellurobismuthite **65**
Tennantite **60**
Tenorite **94**
Tetradymite **65**
Tetrahedrite **60 — 61**, *54 — 55*
Thailandite 344
Thomsonite **311**, *359*
Thulite **268**
Thuringite **297**
Tiger's-eye **112**, 127, *6, 135, 136*
Tirolite **219**
Titanite **259 — 260**, *294*
Topaz 21, 27, 30, **251 — 255**, 280, 281, 296, *285 — 291*
Topazolite s. Demantoid
Torbernite **231**, *263*
Tourmaline 31, 129, 174, 216, 223, 252, 262, 269, **278 — 281**, 285, 296, *314 — 318*
Travertine **194**, *219*
Tremolite **287**
Tripolite 162
Trona **190**, *207*
Turquoise 16, 17, **229**, *2, 4, 260 — 261*

Uralite 285
Uraninite 36, **181 — 182**, 212, *16, 196 — 198*
Uranopilite 212
Utahlite **228**
Uvarovite **241**

Valchovite **336**, *380*
Valencianite **323**
Valentinite **103**, *103*
Vanadinite **221**, *250*
Variscite 216, **228**, 232, *259*
Velvet ore s. Malachite
Verdelite **279**, 281
Vesignieite **219**
Vesuvianite 240, **262 — 264**, *296 — 297*
Vilyuyite **262**
Vitreous schorl s. Axinite
Vivianite 216, **226**, *257*
Vltavine **342**, 343, 344, *387, 388, 390*

Wad **172**, 288, *185*
Wardite 228, **232**, *264*
Wavellite **230**, *262*
Whewellite **331**, *378*